Yuri Hakopian

COMPUTATION OF THE MOORE-PENROSE INVERSE FOR BIDIAGONAL MATRICES

PREFACE

The Moore-Penrose inverse is the most popular type of matrix generalized inverses, or pseudoinverses, which has many applications both in matrix theory and numerical linear algebra. A common use of the Moore-Penrose inverse is to compute the least squares solution to the systems of linear algebraic equations.

It is well known that the Moore-Penrose inverse can be found via singular value decomposition. In this regard, there is the most effective algorithm which consists of two stages. In the first stage, with the help of Householder reflections, the initial matrix is reduced to an upper bidiagonal form (the Golub-Kahan bidiagonalization algorithm). The second stage is known in scientific literature as the Golub-Reinsch algorithm. This is an iterative procedure which with the help of Givens rotations generates a sequence of bidiagonal matrices converging to a diagonal form. Acting in this way, an iterative approximation to the singular value decomposition of the bidiagonal matrix is obtained.

The principal intention of the present research monograph booklet is to develop a method which can be considered as an alternative to the Golub-Reinsch iterative algorithm. Realizing the approach proposed in the study, the following two main results were achieved. First, we obtain explicit expressions for the entries of the Moore-Penrose inverse of upper bidigonal matrices. Secondly, based on the closed form formulae, we get a finite numerical algorithm of optimal order of computational complexity. Thus, we can compute the Moore-Penrose inverse of an upper bidiagonal matrix without using the singular value decomposition.

This book is intended for scientists interested in linear algebra and the theory of generalized inverses. We hope that the work should also be useful for research workers in numerical analysis and computational practitioners.

Yerevan State University,
Armenia

Yuri Hakopian

CONTENTS

Chapter 1. **INTRODUCTION** **4**

1.1 The Moore-Penrose inverse 4

1.2 Computing the Moore-Penrose inverse using the bidiagonalization of the initial matrix 5

1.3 The purpose of the study 7

Chapter 2. **A CLOSED FORM SOLUTION TO THE MOORE-PENROSE INVERSE** **8**

2.1 Partitioning an initial matrix into blocks 8

2.2 A way of computing the Moore-Penrose inverse 10

2.3 Calculation of the block Z_1 13

2.4 Inversion of a pattern matrix $L(\varepsilon)$ 24

2.5 Calculation of the pattern matrices Z and H: type 1 31

2.6 Calculation of the pattern matrices Z and H: type 2 36

2.7 Calculation of the pattern matrices Z and H: type 3 40

2.8 Formulae for the entries of the Moore-Penrose inverse 42

Chapter 3. **AN ALGORITHM TO COMPUTE THE MOORE-PENROSE INVERSE** **47**

3.1 An algorithm of computing the block Z_1 47

3.2 Algorithms for the pattern matrices Z and H 49

3.3 A general computational procedure 51

Bibliography 54

Chapter 1

INTRODUCTION

1.1 The Moore-Penrose inverse

In matrix theory, the term *generalized inverse* or *pseudoinverse* of a matrix A is a generalization of the ordinary notion of the matrix inverse. It is well known that if the matrix A is square and nonsingular, then it has a unique inverse denoted by A^{-1}, such that

$$A^{-1}A = AA^{-1} = I,$$

where I is the identity matrix. If, however, the matrix is singular, or rectangular, then it is not invertable and the symbol A^{-1} is meaningless.

For the first time in the mathematical literature, the concept of a pseudoinverse was introduced by E.I. Fredholm in [5], where generalized inverses of integral operators were investigated. Later, in 1912, A. Hurwitz [13] obtained an algebraic formulation for pseudoinverse of Fredholm integral operators.

In 1920, exploring the pseudoinverses of differential and integral operators, E. Moore [16] introduced the concept of pseudoinverse matrix. However, in the subsequent years continuation of the works in this direction was not followed. The interest in the pseudoinverse matrices, after almost the thirty-year break, reappeared again in the 50s of the last century, in connection with the activation of studies on the least squares method. Here we should mention the papers [3, 4] of A. Bjerhammar, in which the author newly re-discovered the Moore's pseudoinverse and also pointed to the link between pseudoinverse and solving systems of linear algebraic equations.

In 1955 R. Penrose [17] summarized the results of Bjerhammar and showed that for every $m \times n$ matrix A there is a unique $n \times m$ matrix X satisfying the following four equations (so-called *Penrose equations*):

$$\begin{aligned} &AXA = A, \\ &XAX = X, \\ &(AX)^* = AX, \\ &(XA)^* = XA, \end{aligned}$$

where the symbol * denotes the conjugate transpose. Since this generalized inverse X had previously been studied (though defined in a different way) by E. Moore, now it is commonly known as the *Moore-Penrose inverse* and is usually denoted by A^+. It is clear that if A is nonsingular then A^+ coincides with A^{-1}.

The Moore-Penrose inverse has numerous applications in the least squares problems, regressive analysis, linear programming and etc. (see [1], for instance). A survey of the theory of generalized inverses with applications in many areas can be found in extensive monograph [2] of A. Ben-Israel and T.N.E. Greville.

There is a well-known formula for the Moore-Penrose inverse obtained by *singular value decomposition* (abbreviated SVD) of the matrix (see [2, 7], for instance).

The singular value decomposition of an $m \times n$ matrix A with rank r is a factorization of the form

$$A = UDV^*, \tag{1.1.1}$$

where U is an $m \times m$ unitary matrix, $D = diag\,[\sigma_1, \sigma_2, \ldots, \sigma_r]$ is an $m \times n$ diagonal matrix, and V is an $n \times n$ unitary matrix. The diagonal entries $\sigma_1 \geq \sigma_2 \geq \ldots \geq \sigma_r > 0$ of D are known as the *singular values* of the matrix A.

Having the factorization (1.1.1), the Moore-Penrose inverse can be written as

$$A^+ = VD^+U^*, \tag{1.1.2}$$

where the pseudoinverse $D^+ = diag\,[\sigma_1^{-1}, \sigma_2^{-1}, \ldots, \sigma_r^{-1}]$ is the $n \times m$ diagonal matrix.

If the matrix A is real, the matrices U and V in (1.1.1) are orthogonal and the formula (1.1.2) takes the form

$$A^+ = VD^+U^T. \tag{1.1.3}$$

Throughout this paper we will confine ourselves to the real matrices.

1.2 Computing the Moore-Penrose inverse using the bidiagonalization of the initial matrix

Let A be an $m \times n$ real matrix and $m \geq n$. The most effective procedure to compute the Moore-Penrose inverse involves two main stages [7].

Stage 1. *Matrix reduction to the bidiagonal form.*

At this stage the initial matrix A by means of an $m \times m$ orthogonal matrix U and an $n \times n$ orthogonal matrix V is reduced to an upper bidiagonal form:

$$B = U^TAV = \left[\begin{array}{ccccc} d_1 & b_1 & 0 & \ldots & 0 \\ 0 & d_2 & b_2 & \ldots & 0 \\ \vdots & \ddots & \ddots & \ddots & \vdots \\ 0 & \ldots & 0 & d_{n-1} & b_{n-1} \\ 0 & \ldots & 0 & 0 & d_n \\ \hline & & & & \\ & & 0 & & \end{array}\right]. \tag{1.2.1}$$

The matrices U and V are obtained by a series of Householder reflections, alternatively applied from the left and right:

$$U = U_1 \dots U_n \quad \text{and} \quad V = V_1 \dots V_{n-2}\,.$$

The matrix U_j, $1 \le j \le n$ produces zeros in the jth column while the matrix V_j, $1 \le j \le n-2$ nullifies the entries in the jth row. The computational process is known as *Golub-Kahan bidiagonalization* [6]. The algorithm, to compute the matrix B as well as the orthogonal matrices U and V, requires $4mn(m+n) - 4n^3/3$ arithmetical operations [7].

From (1.2.1) we have

$$A = UBV^T.$$

Hence, it follows that

$$A^+ = VB^+U^T. \tag{1.2.2}$$

Thereby the problem is reduced to the Moore-Penrose inversion of the bidiagonal matrix B.

Stage 2. *Golub-Reinsch SVD iterative algorithm.*

Once the bidiagonalization of the matrix A has been achieved, the next task is to zero the superdiagonal entries in the matrix B. To this purpose the *Golub-Reinsch algorithm* is implemented [8]. This iterative process generates a sequence of matrices

$$B^{(k+1)} = S^{(k)}B^{(k)}P^{(k)},\ k = 0, 1, \dots\ ;\ B^{(0)} = B, \tag{1.2.3}$$

which converge to a diagonal matrix. The orthogonal matrices $S^{(k)}$ and $P^{(k)}$ are constructed as a product of Givens rotations. Notice that on each iterative step the matrices $B^{(k)}$ remain bidiagonal. The iterations (1.2.3) are performed unless and until the matrix

$$\hat{D} \equiv S^{(t)} \dots S^{(1)}S^{(0)}BP^{(0)}P^{(1)} \dots P^{(t)} \tag{1.2.4}$$

become practically diagonal. Thus,

$$\hat{D} = SBP, \tag{1.2.5}$$

where, according to (1.2.4),

$$S \equiv S^{(t)} \dots S^{(1)}S^{(0)}, \quad P \equiv P^{(0)}P^{(1)} \dots P^{(t)}.$$

Neglecting small superdiagonal entries of the matrix $\hat{D}$, from (1.2.5) we get a diagonal matrix

$$D \approx SBP.$$

Hereby we find an approximate factorization

$$B \approx S^T DP^T, \tag{1.2.6}$$

which is considered as an approximation to the singular value decomposition of the bidiagonal matrix B. As a result, according to the formula (1.1.3), from (1.2.6) we get

$$B^+ \approx PD^+S.$$

Finally, to compute the matrix A^+, it should be referred to the factored form (1.2.2).

We have considered rectangular $m \times n$ matrices A, where $m \geq n$. The case $m < n$ is reduced to the discussed one since according to the known properties of the pseudoinverse we have

$$A^+ = ((A^T)^T)^+ = ((A^T)^+)^T.$$

1.3 The purpose of the study

The objective of the present work is to develop a method which allows to deduce formulae for the entries of the Moore-Penrose inverse of upper bidiagonal matrices. Obtained closed form solution to the Moore-Penrose inversion may be considered as an alternative to sufficiently labour-consuming Golub-Reinsch iterative procedure, briefly described in the Stage 2 of the previous section. Moreover, explicit expressions for the entries of the Moore-Penrose inverse lead to fairly simple finite numerical algorithm, with optimal volume of computational expenditures. Selected results of this work were published in the author's articles [9, 10, 11, 12].

To move further, let us point out that we will consider square singular upper bidiagonal matrices, i.e. when $m = n$. This is sufficient for our purpose. Indeed, if $m > n$ then from (1.2.1) we have the block structure

$$B = \begin{bmatrix} \hat{B} \\ 0 \end{bmatrix},$$

where $\hat{B}$ is a square upper bidiagonal matrix. It can be readily seen that

$$B^+ = \begin{bmatrix} \hat{B}^+ & 0^T \end{bmatrix}.$$

Chapter 2

A CLOSED FORM SOLUTION TO THE MOORE-PENROSE INVERSE

In this part of the study, we derive explicit formulae for the entries of the Moore-Penrose inverse of singular upper bidiagonal matrices.

2.1 Partitioning an initial matrix into blocks

Let us consider an $n \times n$ real upper bidiagonal matrix

$$A = \begin{bmatrix} d_1 & b_1 & & & \\ & d_2 & b_2 & 0 & \\ & & \ddots & \ddots & \\ & 0 & & d_{n-1} & b_{n-1} \\ & & & & d_n \end{bmatrix}. \qquad (2.1.1)$$

We assume that the matrix A is singular, i.e. $d_1 d_2 \ldots d_n = 0$. Next, we assume that

$$b_1, b_2, \ldots, b_{n-1} \neq 0. \qquad (2.1.2)$$

Otherwise, if some of superdiagonal entries of the matrix A are equal to zero, the problem of computing the Moore-Penrose inverse is decomposed into several similar problems for bidiagonal matrices of lower order.

To compute the Moore-Perrose inverse A^+ of the matrix A from (2.1.1), we represent it in block form

$$A = \begin{bmatrix} A_1 & B_1 & & & \\ & A_2 & B_2 & 0 & \\ & & \ddots & \ddots & \\ & 0 & & A_{p-1} & B_{p-1} \\ & & & & A_p \end{bmatrix} \qquad (2.1.3)$$

with diagonal blocks A_k, $k = 1, 2, \ldots, p$ of the size $n_k \times n_k$ and superdiagonal blocks B_k, $k = 1, 2, \ldots, p-1$ of the size $n_k \times n_{k+1}$, where $n_1 + n_2 + \cdots + n_p = n$. Below we describe the partitioning process. It is carried out proceeding from the location of the zeros on the main diagonal of the matrix A.

Let us start with the first diagonal block A_1. As a basis we charge the following two alternative options:

i) $\underbrace{d_1, d_2, \ldots, d_{n_1-1}, 0}_{n_1}, \ldots,$ where $d_i \neq 0,\ 1 \le i \le n_1 - 1,$

ii) $\underbrace{0, 0, \ldots, 0}_{n_1}, d_{n_1+1}, \ldots,$ where $d_{n_1+1} \neq 0.$

Thereby the diagonal block A_1 of the size $n_1 \times n_1$ is uniquely determined. The result is an intermediate partition

$$A = \left[\begin{array}{c|c} A_1 & \\ \hline & \tilde{A} \end{array}\right]$$

of the matrix A. Next, we apply the same procedure to the submatrix $\tilde{A}$, etc. It is easy to see that the described partition (2.1.3) of the matrix A is unique.

Thus, in the partition (2.1.3) the diagonal blocks A_k, $1 \le k \le p$ are the blocks of one of the following three types:

type 1 : all diagonal entries of the block are nonzero;
type 2 : all diagonal entries of the block, except the last one, are nonzero;
type 3 : all diagonal entries of the block are zero.

In Fig. 2.1.1 we schematically show selected diagonal blocks (the mark $\times$ stands for a nonzero entry).

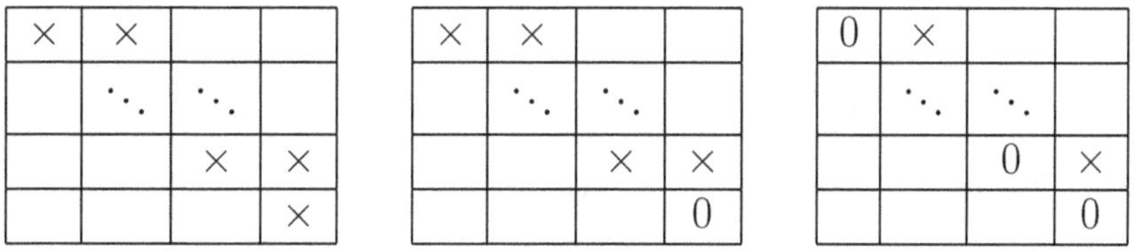

type 1 type 2 type 3

Fig. 2.1.1. The types of diagonal blocks.

Remark 2.1.1 *We emphasize two important features of the partition. Firstly, two blocks of type 3 are not diagonally adjacent, and secondly, as a block of type 1 can be only the last block A_p.*

By virtue of the partitioning rule, the blocks B_k, $k = 1, 2, \ldots, p-1$ have the following structure:

$$B_k = \begin{bmatrix} 0 & 0 & \ldots & 0 \\ \vdots & \vdots & \ldots & \vdots \\ 0 & 0 & \ldots & 0 \\ \Delta_k & 0 & \ldots & 0 \end{bmatrix}, \qquad \Delta_k \equiv b_{n_1+n_2+\cdots+n_k}. \tag{2.1.4}$$

2.2 A way of computing the Moore-Penrose inverse

Here we outline the path of finding the Moore-Penrose inverse based on the block structure (2.1.3) of the matrix A. For this purpose we make use of well-known equality

$$A^{+} = \lim_{\varepsilon\to+0}(A^TA+\varepsilon I)^{-1}A^T, \tag{2.2.1}$$

where I is the identity matrix (see [2, 7], for example). Notice that the equality holds true for any real matrix.

Proceeding from (2.1.3), we have

$$A^TA+\varepsilon I = \begin{bmatrix} L_1(\varepsilon) & A_1^TB_1 & & & \\ B_1^TA_1 & L_2(\varepsilon) & A_2^TB_2 & 0 & \\ & \ddots & \ddots & \ddots & \\ & 0 & B_{p-2}^TA_{p-2} & L_{p-1}(\varepsilon) & A_{p-1}^TB_{p-1} \\ & & & B_{p-1}^TA_{p-1} & L_p(\varepsilon) \end{bmatrix},$$

where

$$L_1(\varepsilon) = A_1^TA_1 + \varepsilon I_1\,, \tag{2.2.2}$$

$$L_k(\varepsilon) = A_k^TA_k + B_{k-1}^TB_{k-1} + \varepsilon I_k\,, \quad k = 2,3,\dots,p \tag{2.2.3}$$

(here and below I_k stands for the identity matrix of order n_k). Since each A_k, $1\le k\le p-1$ is a block of type 2 or type 3, one may easily make sure that $A_k^TB_k = 0$, $1\le k\le p-1$. Consequently, $A^TA+\varepsilon I$ is a block diagonal matrix:

$$A^TA+\varepsilon I = \begin{bmatrix} L_1(\varepsilon) & & & & \\ & L_2(\varepsilon) & 0 & & \\ & 0 & \ddots & & \\ & & & L_{p-1}(\varepsilon) & \\ & & & & L_p(\varepsilon) \end{bmatrix}. \tag{2.2.4}$$

Having the block forms (2.1.3) and (2.2.4), we obtain

$$(A^TA+\varepsilon I)^{-1}A^T =$$

$$\begin{bmatrix} L_1(\varepsilon)^{-1}A_1^T & & & & \\ L_2(\varepsilon)^{-1}B_1^T & L_2(\varepsilon)^{-1}A_2^T & & 0 & \\ \ddots & \ddots & & & \\ 0 & L_{p-1}(\varepsilon)^{-1}B_{p-2}^T & L_{p-1}(\varepsilon)^{-1}A_{p-1}^T & & \\ & & L_p(\varepsilon)^{-1}B_{p-1}^T & L_p(\varepsilon)^{-1}A_p^T & \end{bmatrix}.$$

Hence, according to the equality (2.2.1) we find

$$A^{+} = \begin{bmatrix} Z_1 & & & & \\ H_2 & Z_2 & 0 & & \\ & \ddots & \ddots & & \\ & 0 & H_{p-1} & Z_{p-1} & \\ & & & H_p & Z_p \end{bmatrix}, \tag{2.2.5}$$

where

$$Z_k = \lim_{\varepsilon \to +0} L_k(\varepsilon)^{-1} A_k^T, \quad k = 1, 2, \ldots, p \tag{2.2.6}$$

and

$$H_k = \lim_{\varepsilon \to +0} L_k(\varepsilon)^{-1} B_{k-1}^T, \quad k = 2, 3, \ldots, p. \tag{2.2.7}$$

Let us make a remark concerning the blocks Z_k defined in (2.2.6). For $k = 1$, as follows from (2.2.1) and (2.2.2), we have

$$Z_1 = \lim_{\varepsilon \to +0} (A_1^T A_1 + \varepsilon I_1)^{-1} A_1^T = A_1^+. \tag{2.2.8}$$

However, we can not claim the same with respect to the blocks Z_k, $k \geq 2$, due to (2.2.3).

Further, as seen from definitions (2.2.6) and (2.2.7) of the blocks Z_k and H_k, respectively, the main problem here is to invert the matrices $L_k(\varepsilon)$. First of all let us reveal the structure of these matrices.

As already stated above, the first diagonal block A_1 can be a block of type 2 or type 3. If the block is of type 2, i.e.

$$A_1 = \begin{bmatrix} d_1 & b_1 & & & \\ & d_2 & b_2 & 0 & \\ & & \ddots & \ddots & \\ & 0 & & d_{n_1-1} & b_{n_1-1} \\ & & & & 0 \end{bmatrix}, \tag{2.2.9}$$

(remind that $d_{n_1} = 0$) then

$$L_1(\varepsilon) = \begin{bmatrix} d_1^2 + \varepsilon & b_1 d_1 & & & \\ b_1 d_1 & d_2^2 + b_1^2 + \varepsilon & b_2 d_2 & 0 & \\ & \ddots & \ddots & \ddots & \\ & 0 & b_{n_1-2} d_{n_1-2} & d_{n_1-1}^2 + b_{n_1-2}^2 + \varepsilon & b_{n_1-1} d_{n_1-1} \\ & & & b_{n_1-1} d_{n_1-1} & b_{n_1-1}^2 + \varepsilon \end{bmatrix}. \tag{2.2.10}$$

The case where A_1 is a block of type 3 is much easier to examine; we will consider it below, in corresponding place.

For $k \geq 2$, to simplify the record of formulae, we introduce a local numbering of the entries in blocks A_k, namely

$$A_k = \begin{bmatrix} d_1^{(k)} & b_1^{(k)} & & & \\ & d_2^{(k)} & b_2^{(k)} & 0 & \\ & & \ddots & \ddots & \\ & 0 & & d_{n_k-1}^{(k)} & b_{n_k-1}^{(k)} \\ & & & & d_{n_k}^{(k)} \end{bmatrix}, \tag{2.2.11}$$

where, according to the form (2.1.1) of the matrix A, we have

$$\begin{aligned} d_i^{(k)} &= d_{n_1+\cdots+n_{k-1}+i}\,, \quad i = 1, 2, \ldots, n_k\,, \\ b_i^{(k)} &= b_{n_1+\cdots+n_{k-1}+i}\,, \quad i = 1, 2, \ldots, n_k - 1\,. \end{aligned} \tag{2.2.12}$$

If A_k is a block of type 2 or type 3, then $d_{n_k}^{(k)} = 0$. Thereby, as follows from (2.2.3),

$$L_k(\varepsilon) =$$

$$= \begin{bmatrix} {d_1^{(k)}}^2 + \Delta_{k-1}^2 + \varepsilon & b_1^{(k)} d_1^{(k)} & & & \\ b_1^{(k)} d_1^{(k)} & {d_2^{(k)}}^2 + {b_1^{(k)}}^2 + \varepsilon & b_2^{(k)} d_2^{(k)} & 0 & \\ & \ddots & \ddots & \ddots & \\ & 0 & b_{n_k-2}^{(k)} d_{n_k-2}^{(k)} & {d_{n_k-1}^{(k)}}^2 + {b_{n_k-2}^{(k)}}^2 + \varepsilon & b_{n_k-1}^{(k)} d_{n_k-1}^{(k)} \\ & & & b_{n_k-1}^{(k)} d_{n_k-1}^{(k)} & {d_{n_k}^{(k)}}^2 + {b_{n_k-1}^{(k)}}^2 + \varepsilon \end{bmatrix}, \tag{2.2.13}$$

where $\Delta_{k-1} = b_{n_1+\cdots+n_{k-1}}$ (see (2.1.4)).

Thus, $L_k(\varepsilon)$ are tridiagonal matrices with special structure. To invert these matrices, let us take advantage of an algorithm developed in [15]. Consider a nonsingular symmetric tridiagonal matrix

$$C = \begin{bmatrix} c_{11} & c_{12} & & & \\ c_{21} & c_{22} & c_{23} & & 0 \\ & \ddots & \ddots & \ddots & \\ 0 & & c_{m-1\,m-2} & c_{m-1\,m-1} & c_{m-1\,m} \\ & & & c_{m\,m-1} & c_{m\,m} \end{bmatrix}, \tag{2.2.14}$$

where $c_{i\,i-1} = c_{i-1\,i} \neq 0\,, i = 2, 3, \ldots, m$. We assume that $m \geq 2$. Referring to [15], the matrix $C^{-1} = [x_{ij}]_{m\times m}$ can be obtained by the following computational procedure.

Procedure 3d/inv $(C \Rightarrow C^{-1})$

1. Compute the quantities $f_i\,(i = 2, 3, \ldots, m)$, $g_i\,(i = 2, 3, \ldots, m-1)$ and $h_i\,(i = 1, 2, \ldots, m-1)$:

$$f_i = \frac{c_{ii}}{c_{i\,i-1}}\,, \quad g_i = \frac{c_{i\,i+1}}{c_{i\,i-1}}\,, \quad h_i = \frac{c_{ii}}{c_{i\,i+1}}\,. \tag{2.2.15}$$

Note: if $m = 2$, then the quantities g_i are not introduced.

2. Compute recursively the quantities $\mu_i\,(i = 1, 2, \ldots, m)$:

$$\begin{aligned} &\mu_m = 1\,, \; \mu_{m-1} = -f_m\,, \\ &\mu_i = -f_{i+1}\mu_{i+1} - g_{i+1}\mu_{i+2}\,, \; i = m-2, m-3, \ldots, 1. \end{aligned} \tag{2.2.16}$$

3. Compute recursively the quantities $\nu_i\,(i = 1, 2, \ldots, m)$:

$$\begin{aligned} &\nu_1 = 1\,, \; \nu_2 = -h_1\,, \\ &\nu_i = -h_{i-1}\nu_{i-1} - \frac{1}{g_{i-1}}\nu_{i-2}\,, \; i = 3, 4, \ldots, m. \end{aligned} \tag{2.2.17}$$

4. Compute the quantity

$$t = (c_{11}\mu_1 + c_{12}\mu_2)^{-1}. \tag{2.2.18}$$

Note: if C is nonsingular matrix, then $c_{11}\mu_1 + c_{12}\mu_2 \neq 0$ [15].

5. The entries of the lower triangular part of the matrix C^{-1} are computed:

$$x_{ij} = \mu_i \nu_j t,\ i = j, j+1, \ldots, m; \quad j = 1, 2, \ldots, m. \tag{2.2.19}$$

6. The entries of the upper triangular part of the matrix C^{-1} are computed:

$$x_{ij} = \mu_j \nu_i t,\ i = 1, 2, \ldots, j-1; \quad j = 2, 3, \ldots, m. \tag{2.2.20}$$

Note: since the matrix C^{-1} is also symmetric, then $x_{ij} = x_{ji}$.

End procedure

Proposed way to obtain the blocks Z_k and H_k in the block representation (2.2.5) of the matrix A^+ consists in the following. Finding first the inverse matrices $L_k(\varepsilon)^{-1}$, the entries of the matrices $L_k(\varepsilon)^{-1}A_k^T$ and $L_k(\varepsilon)^{-1}B_{k-1}^T$ are calculated and a character of their dependence on the parameter ε is revealed. Then, according to the equalities (2.2.6) and (2.2.7), passing to the limit when $\varepsilon \to +0$, we arrive to a closed form expressions for the entries of the Moore-Penrose inverse A^+.

Remark 2.2.1 *In expressions below we will deal with products and sums of different quantities. When recording*

$$\prod_{i=m}^{n} a_i\,, \quad \sum_{i=m}^{n} a_i\,,$$

it is usually assumed that $m \leq n$. If $m > n$, as it is customary in computations, the evaluated product is assumed to be unity, and the sum is equal to zero.

2.3 Calculation of the block Z_1

In this section we deduce formulae for the entries of the first diagonal block

$$Z_1 = [z_{ij}^{(1)}]_{n_1 \times n_1}$$

in the block representation (2.2.5) of the matrix A^+. According to (2.2.8), $Z_1 = A_1^+$.

We start with the case, when A_1 is a block of type 2.

Let us consider as the matrix C our tridiagonal matrix $L_1(\varepsilon)$. Comparing the records (2.2.10), (2.2.14) of these matrices and setting $m = n_1$, we have

$$c_{ii} = d_i^2 + b_{i-1}^2 + \varepsilon\,,\ i = 1, 2, \ldots, n_1 \tag{2.3.1}$$

(in order to unify records of formulae, we set $b_0 = 0$) and

$$c_{i\,i+1} = b_i d_i\,,\ i = 1, 2, \ldots, n_1 - 1; \quad c_{i\,i-1} = b_{i-1} d_{i-1}\,,\ i = 2, 3, \ldots, n_1. \tag{2.3.2}$$

In accordance with our plan, let us carry out a more detailed analysis of the quantities successively computed in the procedure **3d/inv** from the Section 2.2.

Consider first the quantities f_i, g_i and h_i which were introduced in (2.2.15). Using the expressions (2.3.1) and (2.3.2), we get

$$f_i = \overset{\circ}{f}_i + \alpha_i \varepsilon\,,\ i = 2,3,\ldots,n_1, \text{ where } \overset{\circ}{f}_i = \frac{d_i^2 + b_{i-1}^2}{b_{i-1}d_{i-1}}\,,\ \alpha_i = \frac{1}{b_{i-1}d_{i-1}}\,; \tag{2.3.3}$$

$$g_i = \frac{b_i d_i}{b_{i-1}d_{i-1}}\,,\ i = 2,3,\ldots,n_1 - 1; \tag{2.3.4}$$

$$h_i = \overset{\circ}{h}_i + \beta_i \varepsilon\,,\ i = 1,2,\ldots,n_1 - 1, \text{ where } \overset{\circ}{h}_i = \frac{d_i^2 + b_{i-1}^2}{b_i d_i}\,,\ \beta_i = \frac{1}{b_i d_i}\,. \tag{2.3.5}$$

Next, go to the quantities μ_i and ν_i recursively defined in (2.2.16) and (2.2.17), respectively.

Lemma 2.3.1 *The quantities μ_i are represented as*

$$\begin{aligned} &\mu_{n_1} = \overset{\circ}{\mu}_{n_1} + \gamma_{n_1}\varepsilon\,, \quad \mu_{n_1-1} = \overset{\circ}{\mu}_{n_1-1} + \gamma_{n_1-1}\varepsilon\,, \\ &\mu_i = \overset{\circ}{\mu}_i + \gamma_i\varepsilon + O(\varepsilon^2)\,, \quad 1 \le i \le n_1 - 2\,, \end{aligned} \tag{2.3.6}$$

where the quantities $\overset{\circ}{\mu}_i$ and γ_i satisfy the following recurrence relations:

$$\begin{aligned} &\overset{\circ}{\mu}_{n_1} = 1\,,\ \overset{\circ}{\mu}_{n_1-1} = -\overset{\circ}{f}_{n_1}\,, \\ &\overset{\circ}{\mu}_i = -\overset{\circ}{f}_{i+1}\overset{\circ}{\mu}_{i+1} - g_{i+1}\overset{\circ}{\mu}_{i+2}\,,\ i = n_1 - 2, n_1 - 3, \ldots, 1 \end{aligned} \tag{2.3.7}$$

and

$$\begin{aligned} &\gamma_{n_1} = 0\,,\ \gamma_{n_1-1} = -\alpha_{n_1}\,, \\ &\gamma_i = -\overset{\circ}{f}_{i+1}\gamma_{i+1} - g_{i+1}\gamma_{i+2} - \alpha_{i+1}\overset{\circ}{\mu}_{i+1}\,,\ i = n_1 - 2, n_1 - 3, \ldots, 1. \end{aligned} \tag{2.3.8}$$

Proof. Since $\mu_{n_1} = 1$, then in (2.3.6) we set $\overset{\circ}{\mu}_{n_1} = 1$, $\gamma_{n_1} = 0$. Further, $\mu_{n_1-1} = -f_{n_1}$ (see (2.2.16)). According to the expressions (2.3.3), we have $f_{n_1} = \overset{\circ}{f}_{n_1} + \alpha_{n_1}\varepsilon$. Therefore, in the representation (2.3.6) we set $\overset{\circ}{\mu}_{n_1-1} = -\overset{\circ}{f}_{n_1}$, $\gamma_{n_1-1} = -\alpha_{n_1}$.

Required representations, for the indices in the range $1 \le i \le n_1 - 2$, can be readily derived by induction from the relations (2.2.16), using expressions (2.3.3). Indeed, having done simple transformations as follows

$$\begin{aligned} \mu_i &= -f_{i+1}\mu_{i+1} - g_{i+1}\mu_{i+2} \\ &= -(\overset{\circ}{f}_{i+1} + \alpha_{i+1}\varepsilon)(\overset{\circ}{\mu}_{i+1} + \gamma_{i+1}\varepsilon + O(\varepsilon^2)) - g_{i+1}(\overset{\circ}{\mu}_{i+2} + \gamma_{i+2}\varepsilon + O(\varepsilon^2)) \\ &= (-\overset{\circ}{f}_{i+1}\overset{\circ}{\mu}_{i+1} - g_{i+1}\overset{\circ}{\mu}_{i+2}) + (-\overset{\circ}{f}_{i+1}\gamma_{i+1} - g_{i+1}\gamma_{i+2} - \alpha_{i+1}\overset{\circ}{\mu}_{i+1})\varepsilon + O(\varepsilon^2)\,, \end{aligned}$$

we get (2.3.6), as well as recurrence relations (2.3.7) and (2.3.8). □

The quantities $\overset{\circ}{\mu}_i$ computed by the recursion (2.3.7) may be represented in closed form. Let us introduce the following notation:

$$r_s^{(1)} \equiv \frac{b_s}{d_s}, \qquad s = 1, 2, \ldots, n_1 - 1 . \tag{2.3.9}$$

Additionally, we set $r_0^{(1)} = r_{n_1}^{(1)} = 1$.

Lemma 2.3.2 *The quantities $\overset{\circ}{\mu}_i$ can be written in the form*

$$\overset{\circ}{\mu}_i = (-1)^{n_1 - i} \prod_{s=i}^{n_1-1} r_s^{(1)}, \quad i = 1, 2, \ldots, n_1. \tag{2.3.10}$$

Proof. Firstly, the value $\overset{\circ}{\mu}_{n_1} = 1$ conforms to the record (2.3.10). Then, in accordance with (2.3.3) and (2.3.7),

$$\overset{\circ}{\mu}_{n_1-1} = -\overset{\circ}{f}_{n_1} = -\frac{b_{n_1-1}}{d_{n_1-1}} = -r_{n_1-1}^{(1)} .$$

Further reasoning is carried out by induction. Using the expressions (2.3.3) and (2.3.4), proceeding from (2.3.7) we obtain

$$\begin{aligned} \overset{\circ}{\mu}_i &= -\frac{d_{i+1}^2 + b_i^2}{b_i d_i}(-1)^{n_1-i-1} \prod_{s=i+1}^{n_1-1} r_s^{(1)} - \frac{b_{i+1}d_{i+1}}{b_i d_i}(-1)^{n_1-i-2} \prod_{s=i+2}^{n_1-1} r_s^{(1)} \\ &= (-1)^{n_1-i} \prod_{s=i+2}^{n_1-1} r_s^{(1)} \left(\frac{d_{i+1}^2 + b_i^2}{b_i d_i} r_{i+1}^{(1)} - \frac{b_{i+1}d_{i+1}}{b_i d_i} \right) = (-1)^{n_1-i} \prod_{s=i}^{n_1-1} r_i^{(1)} , \end{aligned}$$

which completes the proof of the Lemma. □

The next assertion is a simple consequence of the formula (2.3.10).

Corollary 2.3.1 *The following relation holds*

$$\overset{\circ}{\mu}_i = -r_i^{(1)} \overset{\circ}{\mu}_{i+1} , \quad i = 1, 2, \ldots, n_1 - 1 . \tag{2.3.11}$$

A representation similar to (2.3.6) takes place also for the quantities ν_i.

Lemma 2.3.3 *The quantities ν_i are represented as*

$$\begin{aligned} &\nu_1 = \overset{\circ}{\nu}_1 + \delta_1 \varepsilon , \quad \nu_2 = \overset{\circ}{\nu}_2 + \delta_2 \varepsilon , \\ &\nu_i = \overset{\circ}{\nu}_i + \delta_i \varepsilon + O(\varepsilon^2) , \quad 3 \le i \le n_1 , \end{aligned} \tag{2.3.12}$$

where the quantities $\overset{\circ}{\nu}_i$ and δ_i satisfy the following recurrence relations:

$$\begin{aligned} &\overset{\circ}{\nu}_1 = 1 , \ \overset{\circ}{\nu}_2 = -\overset{\circ}{h}_1 , \\ &\overset{\circ}{\nu}_i = -\overset{\circ}{h}_{i-1} \overset{\circ}{\nu}_{i-1} - \frac{1}{g_{i-1}} \overset{\circ}{\nu}_{i-2} , \ i = 3, 4, \ldots, n_1 \end{aligned} \tag{2.3.13}$$

and

$$\delta_1 = 0\,,\ \delta_2 = -\beta_1\,,$$
$$\delta_i = -\overset{\circ}{h}_{i-1}\,\delta_{i-1} - \frac{1}{g_{i-1}}\delta_{i-2} - \beta_{i-1}\,\overset{\circ}{\nu}_{i-1}\,,\ i = 3, 4, \ldots, n_1. \tag{2.3.14}$$

Proof. Since $\nu_1 = 1$, then in (2.3.12) we set $\overset{\circ}{\nu}_1= 1$, $\delta_1 = 0$. Further, $\nu_2 = -h_1$ (see (2.2.17)). According to the expressions (2.3.5), we have $h_1 =\overset{\circ}{h}_1 +\beta_1\varepsilon$. Therefore, in the representation (2.3.12) we set $\overset{\circ}{\nu}_2= -\overset{\circ}{h}_1$, $\delta_2 = -\beta_1$.

Required representations, for the indices in the range $3 \le i \le n_1$, can be readily derived by induction from the relations (2.2.17), using expressions (2.3.5). Indeed, having done simple transformations as follows

$$\begin{aligned}
\nu_i &= -h_{i-1}\nu_{i-1} - \frac{1}{g_{i-1}}\nu_{i-2} \\
&= -(\overset{\circ}{h}_{i-1} +\beta_{i-1}\varepsilon)(\overset{\circ}{\nu}_{i-1} +\delta_{i-1}\varepsilon + O(\varepsilon^2)) - \frac{1}{g_{i-1}}(\overset{\circ}{\nu}_{i-2} +\delta_{i-2}\varepsilon + O(\varepsilon^2)) \\
&= (-\overset{\circ}{h}_{i-1}\overset{\circ}{\nu}_{i-1} - \frac{1}{g_{i-1}}\overset{\circ}{\nu}_{i-2}) + (-\overset{\circ}{h}_{i-1}\,\delta_{i-1} - \frac{1}{g_{i-1}}\delta_{i-2} - \beta_{i-1}\,\overset{\circ}{\nu}_{i-1})\varepsilon + O(\varepsilon^2)\,,
\end{aligned}$$

we get (2.3.12), as well as recurrence relations (2.3.13) and (2.3.14). □

We can write closed form expressions for the quantities $\overset{\circ}{\nu}_i$ as well.

Lemma 2.3.4 *The quantities $\overset{\circ}{\nu}_i$ can be written in the form*

$$\overset{\circ}{\nu}_i= (-1)^{i+1}\prod_{s=1}^{i-1}\frac{1}{r_s^{(1)}}\,, \qquad i = 1, 2, \ldots, n_1. \tag{2.3.15}$$

Proof. The value $\overset{\circ}{\nu}_1= 1$ conforms to the record (2.3.15). Then, in accordance with (2.3.5) and (2.3.13),

$$\overset{\circ}{\nu}_2= -\overset{\circ}{h}_1= -\frac{d_1}{b_1} = -\frac{1}{r_1^{(1)}}\,.$$

Further reasoning is carried out by induction. Taking into account the expressions (2.3.4), (2.3.5) and using (2.3.13), we get

$$\begin{aligned}
\overset{\circ}{\nu}_i &= -\overset{\circ}{h}_{i-1}\overset{\circ}{\nu}_{i-1} - \frac{1}{g_{i-1}}\overset{\circ}{\nu}_{i-2} \\
&= -\frac{d_{i-1}^2 + b_{i-2}^2}{b_{i-1}d_{i-1}}(-1)^i\prod_{s=1}^{i-2}\frac{1}{r_s^{(1)}} - \frac{b_{i-2}d_{i-2}}{b_{i-1}d_{i-1}}(-1)^{i-1}\prod_{s=1}^{i-3}\frac{1}{r_s^{(1)}} \\
&= (-1)^{i+1}\left(\frac{d_{i-1}^2 + b_{i-2}^2}{b_{i-1}d_{i-1}}\frac{1}{r_{i-2}^{(1)}} - \frac{b_{i-2}d_{i-2}}{b_{i-1}d_{i-1}}\right)\prod_{s=1}^{i-3}\frac{1}{r_s^{(1)}} \\
&= (-1)^{i+1}\frac{1}{r_{i-2}^{(1)}}\frac{1}{r_{i-1}^{(1)}}\prod_{s=1}^{i-3}\frac{1}{r_s^{(1)}} = (-1)^{i+1}\prod_{s=1}^{i-1}\frac{1}{r_i^{(1)}}\,,
\end{aligned}$$

which completes the proof of the Lemma. $\sqcap$

The next assertion is a simple consequence of the formula (2.3.15).

Corollary 2.3.2 *The following relation holds*

$$\overset{\circ}{\nu}_{i+1} = -\frac{1}{r_i^{(1)}} \overset{\circ}{\nu}_i, \quad i = 1, 2, \ldots, n_1 - 1. \tag{2.3.16}$$

According to the procedure **3d/inv**, our next task is to derive an expression for the quantity t given in (2.2.18), depending on the parameter ε. Since $c_{11} = d_1^2 + \varepsilon$, $c_{12} = b_1 d_1$ (see (2.3.1) and (2.3.2)), then taking into account the representations (2.3.6) for the quantities μ_i we get

$$\begin{aligned} t &= ((d_1^2 + \varepsilon)(\overset{\circ}{\mu}_1 + \gamma_1 \varepsilon + O(\varepsilon^2)) + b_1 d_1 (\overset{\circ}{\mu}_2 + \gamma_2 \varepsilon + O(\varepsilon^2)))^{-1} \\ &= (d_1^2(\overset{\circ}{\mu}_1 + r_1^{(1)} \overset{\circ}{\mu}_2) + (\overset{\circ}{\mu}_1 + d_1(\gamma_1 d_1 + \gamma_2 b_1))\varepsilon + O(\varepsilon^2))^{-1}. \end{aligned}$$

By virtue of the relation (2.3.11), $\overset{\circ}{\mu}_1 + r_1^{(1)} \overset{\circ}{\mu}_2 = 0$. Thus

$$t = \frac{1}{(\overset{\circ}{\mu}_1 + d_1(\gamma_1 d_1 + \gamma_2 b_1))\varepsilon + O(\varepsilon^2)}. \tag{2.3.17}$$

Having the representations for the quantities μ_i, ν_i and t, by formulae (2.2.19) and (2.2.20) we get the entries of the inverse matrix

$$L_1(\varepsilon)^{-1} = [x_{ij}]_{n_1 \times n_1}.$$

To move further, we introduce a matrix

$$L_1(\varepsilon)^{-1} A_1^T \equiv Y(\varepsilon) = [y_{ij}(\varepsilon)]_{n_1 \times n_1}. \tag{2.3.18}$$

Since $Z_1 = \lim_{\varepsilon \to +0} Y(\varepsilon)$ (see (2.2.8)) then

$$z_{ij}^{(1)} = \lim_{\varepsilon \to +0} y_{ij}(\varepsilon), \quad i, j = 1, 2, \ldots, n_1. \tag{2.3.19}$$

The entries of the last column of the matrix A_1^T are equal to zero (see (2.2.9)). Therefore, $y_{in}(\varepsilon) = 0$, $i = 1, 2, \ldots, n_1$ and by this very fact

$$z_{i n_1}^{(1)} = 0, \quad i = 1, 2, \ldots, n_1. \tag{2.3.20}$$

As follows from (2.3.18), for indices $i = 1, 2, \ldots, n_1$ and $j = 1, 2, \ldots, n_1 - 1$, the entries of the matrix $Y(\varepsilon)$ are calculated by the rule

$$y_{ij}(\varepsilon) = x_{ij} d_j + x_{i\, j+1} b_j. \tag{2.3.21}$$

Subject to the formulae (2.2.19) and (2.2.20), for a fixed index j in the range $1 \le j \le n_1 - 1$ we consider separately two cases: $i = 1, 2, \ldots, j$ and $i = j+1, j+2, \ldots, n_1$.

• Indices $i = 1, 2, \dots, j$.

Taking advantage of the expression (2.2.20) for the entries x_{ij}, from (2.3.21) we can write

$$y_{ij}(\varepsilon) = t\nu_i(\mu_j d_j + \mu_{j+1} b_j). \tag{2.3.22}$$

Then, using the representations (2.3.6) of the quantities μ_i, we have

$$\begin{aligned} \mu_j d_j + \mu_{j+1} b_j &= (\overset{\circ}{\mu}_j + \gamma_j \varepsilon + O(\varepsilon^2)) d_j + (\overset{\circ}{\mu}_{j+1} + \gamma_{j+1} \varepsilon + O(\varepsilon^2)) b_j \\ &= (\overset{\circ}{\mu}_j d_j + \overset{\circ}{\mu}_{j+1} b_j) + (\gamma_j d_j + \gamma_{j+1} b_j)\varepsilon + O(\varepsilon^2). \end{aligned}$$

As follows from the relation (2.3.11),

$$\overset{\circ}{\mu}_j d_j + \overset{\circ}{\mu}_{j+1} b_j = d_j(\overset{\circ}{\mu}_j + r_j^{(1)} \overset{\circ}{\mu}_{j+1}) = 0.$$

Thus,

$$\mu_j d_j + \mu_{j+1} b_j = (\gamma_j d_j + \gamma_{j+1} b_j)\varepsilon + O(\varepsilon^2). \tag{2.3.23}$$

Substituting the expression (2.3.23) as well as the representations (2.3.12) and (2.3.17) of the quantities ν_i and t, respectively, into the right hand side of the equality (2.3.22) yields

$$y_{ij}(\varepsilon) = \frac{\overset{\circ}{\nu}_i (\gamma_j d_j + \gamma_{j+1} b_j) + O(\varepsilon)}{\overset{\circ}{\mu}_1 + d_1(\gamma_1 d_1 + \gamma_2 b_1) + O(\varepsilon)}.$$

By taking limit in the previous equality, according to (2.3.19) we find

$$z_{ij}^{(1)} = \frac{\overset{\circ}{\nu}_i (\gamma_j d_j + \gamma_{j+1} b_j)}{\overset{\circ}{\mu}_1 + d_1(\gamma_1 d_1 + \gamma_2 b_1)}, \qquad i = 1, 2, \dots, j\,.$$

Further, let us introduce the notation

$$u_j \equiv \gamma_j d_j + \gamma_{j+1} b_j, \quad j = 1, 2, \dots, n_1 - 1\,. \tag{2.3.24}$$

Then the entries $z_{ij}^{(1)}$ can be written as follows:

$$z_{ij}^{(1)} = \frac{\overset{\circ}{\nu}_i\, u_j}{q}, \quad i = 1, 2, \dots, j\,, \tag{2.3.25}$$

where

$$q \equiv \overset{\circ}{\mu}_1 + d_1 u_1\,. \tag{2.3.26}$$

Now let us turn to the quantities u_j defined in (2.3.24). For the index $j = n_1 - 1$, using the expressions (2.3.8) and (2.3.3), we have

$$u_{n_1-1} = \gamma_{n_1-1} d_{n_1-1} + \gamma_{n_1} b_{n_1-1} = -\alpha_{n_1} d_{n_1-1} = -\frac{1}{b_{n_1-1}}\,. \tag{2.3.27}$$

For the indices $j = n_1 - 2, n_1 - 3, \dots, 1$, taking advantage of the expressions (2.3.8), (2.3.3) and (2.3.4), we get

$$\begin{aligned}
\gamma_j d_j &= (-\overset{\circ}{f}_{j+1}\,\gamma_{j+1} - g_{j+1}\gamma_{j+2} - \alpha_{j+1}\,\overset{\circ}{\mu}_{j+1})d_j \\
&= -\frac{d_{j+1}^2 + b_j^2}{b_j d_j}\gamma_{j+1}d_j - \frac{b_{j+1}d_{j+1}}{b_j d_j}\gamma_{j+2}d_j - \frac{1}{b_j d_j}\,\overset{\circ}{\mu}_{j+1}\,d_j \\
&= -\gamma_{j+1}b_j - \frac{d_{j+1}^2}{b_j}\gamma_{j+1} - \frac{b_{j+1}d_{j+1}}{b_j}\gamma_{j+2} - \frac{1}{b_j}\,\overset{\circ}{\mu}_{j+1} \\
&= -\gamma_{j+1}b_j - \frac{d_{j+1}}{b_j}(\gamma_{j+1}d_{j+1} + \gamma_{j+2}b_{j+1}) - \frac{1}{b_j}\,\overset{\circ}{\mu}_{j+1}\,.
\end{aligned}$$

Hence,

$$\gamma_j d_j + \gamma_{j+1}b_j = -\frac{d_{j+1}}{b_j}(\gamma_{j+1}d_{j+1} + \gamma_{j+2}b_{j+1}) - \frac{1}{b_j}\,\overset{\circ}{\mu}_{j+1}\,.$$

With a glance to the notation (2.3.24), we arrive at the equality

$$u_j = -\frac{d_{j+1}}{b_j}u_{j+1} - \frac{1}{b_j}\,\overset{\circ}{\mu}_{j+1}\,. \tag{2.3.28}$$

Summarizing the above considerations, on the basis of the obtained equalities (2.3.27) and (2.3.28), we can state that the quantities u_j satisfy the following relations:

$$\begin{aligned}
u_{n_1-1} &= -\frac{1}{b_{n_1-1}}, \\
u_j &= -\frac{d_{j+1}u_{j+1} + \overset{\circ}{\mu}_{j+1}}{b_j}, \qquad j = n_1 - 2, n_1 - 3, \dots, 1\,.
\end{aligned} \tag{2.3.29}$$

The quantities u_j can be represented in closed form as well, namely, the following statement holds.

Lemma 2.3.5 *The quantities u_j are written as*

$$u_j = \frac{(-1)^{n_1-j}}{d_j}\sum_{k=1}^{n_1-j}\left(\prod_{s=j}^{n_1-k}\frac{1}{r_s^{(1)}}\right)\left(\prod_{s=n_1-k+1}^{n_1-1} r_s^{(1)}\right), \quad j = 1, 2, \dots, n_1 - 1\,. \tag{2.3.30}$$

The assertion can be proved by direct substituting the expression (2.3.30) into the relations (2.3.29) and using the expression (2.3.10) for the quantities $\overset{\circ}{\mu}_j$.

As a direct consequence of the expressions (2.3.10) and (2.3.30) we get a closed form expression for the quantity q defined in (2.3.26).

Lemma 2.3.6 *The quantity q is written as*

$$q = (-1)^{n_1-1} \sum_{k=1}^{n_1} \left(\prod_{s=1}^{n_1-k} \frac{1}{r_s^{(1)}} \right) \left(\prod_{s=n_1-k+1}^{n_1-1} r_s^{(1)} \right) . \tag{2.3.31}$$

Finally, let us replace the expressions (2.3.15), (2.3.30) and (2.3.31) of the quantities $\overset{\circ}{\nu}_i$, u_j and q, respectively, into (2.3.25). As a result, we obtain the following closed form expression for the entries of the upper triangular part of the matrix Z_1:

$$z_{ij}^{(1)} = \frac{(-1)^{i+j} \sum\limits_{k=1}^{n_1-j} \left(\prod\limits_{s=j}^{n_1-k} \frac{1}{r_s^{(1)}} \right) \left(\prod\limits_{s=n_1-k+1}^{n_1-1} r_s^{(1)} \right)}{\prod\limits_{s=1}^{i-1} r_s^{(1)} \cdot d_j \sum\limits_{k=1}^{n_1} \left(\prod\limits_{s=1}^{n_1-k} \frac{1}{r_s^{(1)}} \right) \left(\prod\limits_{s=n_1-k+1}^{n_1-1} r_s^{(1)} \right)}, \quad i = 1, 2, \ldots, j . \tag{2.3.32}$$

• Indices $i = j+1, j+2, \ldots, n_1$.
Using the expressions (2.2.19) for the entries x_{ij}, from (2.3.21) we get the equality

$$y_{ij}(\varepsilon) = t\mu_i(\nu_j d_j + \nu_{j+1} b_j). \tag{2.3.33}$$

In accordance with the representations (2.3.12), we have

$$\begin{aligned} \nu_j d_j + \nu_{j+1} b_j &= (\overset{\circ}{\nu}_j + \delta_j \varepsilon + O(\varepsilon^2)) d_j + (\overset{\circ}{\nu}_{j+1} + \delta_{j+1} \varepsilon + O(\varepsilon^2)) b_j \\ &= (\overset{\circ}{\nu}_j d_j + \overset{\circ}{\nu}_{j+1} b_j) + (\delta_j d_j + \delta_{j+1} b_j)\varepsilon + O(\varepsilon^2). \end{aligned}$$

As follows from the relation (2.3.16),

$$\overset{\circ}{\nu}_j d_j + \overset{\circ}{\nu}_{j+1} b_j = d_j(\overset{\circ}{\nu}_j + r_j^{(1)} \overset{\circ}{\nu}_{j+1}) = 0.$$

Thus,

$$\nu_j d_j + \nu_{j+1} b_j = (\delta_j d_j + \delta_{j+1} b_j)\varepsilon + O(\varepsilon^2). \tag{2.3.34}$$

Substituting the expression (2.3.34) as well as the representations (2.3.6) and (2.3.17) of the quantities μ_i and t, respectively, into the right hand side of the equality (2.3.33) yields

$$y_{ij}(\varepsilon) = \frac{\overset{\circ}{\mu}_i (\delta_j d_j + \delta_{j+1} b_j) + O(\varepsilon)}{\overset{\circ}{\mu}_1 + d_1(\gamma_1 d_1 + \gamma_2 b_1) + O(\varepsilon)}.$$

By taking limit in this equality, when $\varepsilon \to +0$, according to (2.3.19) we find

$$z_{ij}^{(1)} = \frac{\overset{\circ}{\mu}_i (\delta_j d_j + \delta_{j+1} b_j)}{\overset{\circ}{\mu}_1 + d_1(\gamma_1 d_1 + \gamma_2 b_1)}, \quad i = j+1, j+2, \ldots, n_1 .$$

Similarly to the previous case, we introduce the notation

$$w_j \equiv \delta_j d_j + \delta_{j+1} b_j, \quad j = 1, 2, \ldots, n_1 - 1. \tag{2.3.35}$$

Then the entries $z_{ij}^{(1)}$ can be written by

$$z_{ij}^{(1)} = \frac{\overset{\circ}{\mu}_i \, w_j}{q}, \quad i = j+1, j+2, \ldots, n_1. \tag{2.3.36}$$

Consider the quantities w_j defined in (2.3.35). For the index $j = 1$, using the expressions (2.3.14) and (2.3.5), we have

$$w_1 = \delta_1 d_1 + \delta_2 b_1 = -\beta_1 b_1 = -\frac{1}{d_1}. \tag{2.3.37}$$

For the indices $j = 2, 3, \ldots, n_1 - 1$, taking advantage of the expressions (2.3.14), (2.3.4) and (2.3.5) yields

$$\begin{aligned}
\delta_{j+1} b_j &= (-\overset{\circ}{h}_j \, \delta_j - \frac{1}{g_j} \delta_{j-1} - \beta_j \, \overset{\circ}{\nu}_j) b_j \\
&= -\frac{d_j^2 + b_{j-1}^2}{b_j d_j} \delta_j b_j - \frac{b_{j-1} d_{j-1}}{b_j d_j} \delta_{j-1} b_j - \frac{1}{b_j d_j} \, \overset{\circ}{\nu}_j \, b_j \\
&= -\delta_j d_j - \frac{b_{j-1}^2}{d_j} \delta_j - \frac{b_{j-1} d_{j-1}}{d_j} \delta_{j-1} - \frac{1}{d_j} \, \overset{\circ}{\nu}_j \\
&= -\delta_j d_j - \frac{b_{j-1}}{d_j} (\delta_j b_{j-1} + \delta_{j-1} d_{j-1}) - \frac{1}{d_j} \, \overset{\circ}{\nu}_j \, .
\end{aligned}$$

Hence,

$$\delta_j d_j + \delta_{j+1} b_j = -\frac{b_{j-1}}{d_j} (\delta_{j-1} d_{j-1} + \delta_j b_{j-1}) - \frac{1}{d_j} \, \overset{\circ}{\nu}_j \, .$$

Taking into account the notation (2.3.35), we get the equality

$$w_j = -\frac{b_{j-1}}{d_j} w_{j-1} - \frac{1}{d_j} \, \overset{\circ}{\nu}_j \, . \tag{2.3.38}$$

Summing up the above considerations, on the basis of the equalities (2.3.37) and (2.3.38), we obtain that the quantities w_j satisfy the following relations:

$$\begin{aligned}
w_1 &= -\frac{1}{d_1}, \\
w_j &= -\frac{b_{j-1} w_{j-1} + \overset{\circ}{\nu}_j}{d_j}, \quad j = 2, 3, \ldots, n_1 - 1.
\end{aligned} \tag{2.3.39}$$

As with u_j, the quantities w_j can be represented in closed form. The following statement holds.

Lemma 2.3.7 *The quantities w_j are written as*

$$w_j = \frac{(-1)^j}{d_j} \sum_{k=1}^{j} \left(\prod_{s=1}^{k-1} \frac{1}{r_s^{(1)}} \right) \left(\prod_{s=k}^{j-1} r_s^{(1)} \right), \quad j = 1, 2, \ldots, n_1 - 1. \tag{2.3.40}$$

The assertion can be proved by direct substitution of the expression (2.3.40) into the relations (2.3.39) and using the expression (2.3.15) for the quantities $\overset{\circ}{\nu}_j$.

Finally, let us replace the expressions (2.3.10), (2.3.40) and (2.3.31) of the quantities $\overset{\circ}{\mu}_i$, w_j and q, respectively, into the equality (2.3.36). Resulting formula for the entries of the lower triangular part of the matrix Z_1 is of the following type:

$$z_{ij}^{(1)} = \frac{(-1)^{i+j+1} \left(\prod_{s=i}^{n_1-1} r_s^{(1)} \right) \sum_{k=1}^{j} \left(\prod_{s=1}^{k-1} \frac{1}{r_s^{(1)}} \right) \left(\prod_{s=k}^{j-1} r_s^{(1)} \right)}{d_j \sum_{k=1}^{n_1} \left(\prod_{s=1}^{n_1-k} \frac{1}{r_s^{(1)}} \right) \left(\prod_{s=n_1-k+1}^{n_1-1} r_s^{(1)} \right)}, \quad i = j+1, j+2, \ldots, n_1. \tag{2.3.41}$$

Thus, in (2.3.20),(2.3.32) and (2.3.41) we have got formulae for the entries of the block Z_1, when A_1 is a block of type 2.

It remains to consider the case when A_1 is a block of type 3. If $n_1 = 1$, then $A_1 = [0]_{1\times 1}$ and, obviously,

$$Z_1 = [0]_{1\times 1} \tag{2.3.42}$$

(see [2], for example). For $n_1 \geq 2$ we have

$$L_1(\varepsilon)^{-1} A_1^T = \begin{bmatrix} 0 & & & & \\ \frac{b_1}{b_1^2+\varepsilon} & 0 & & 0 & \\ & \frac{b_2}{b_2^2+\varepsilon} & 0 & & \\ & & \ddots & \ddots & \\ & 0 & & \frac{b_{n_1-1}}{b_{n_1-1}^2+\varepsilon} & 0 \end{bmatrix}.$$

Passing here to the limit as $\varepsilon \to +0$, according to the equality (2.2.8), we get lower bidiagonal matrix

$$Z_1 = \begin{bmatrix} 0 & & & & \\ b_1^{-1} & 0 & & 0 & \\ & b_2^{-1} & 0 & & \\ & 0 & \ddots & \ddots & \\ & & & b_{n_1-1}^{-1} & 0 \end{bmatrix}. \tag{2.3.43}$$

Combining the above considerations, i.e. having (2.3.20), (2.3.32), (2.3.41), (2.3.42) and (2.3.43), we arrive at the following statement.

Lemma 2.3.8 *The entries of the first diagonal block $Z_1 = [z_{ij}^{(1)}]_{n_1 \times n_1}$ in the block representation (2.2.5) of the matrix A^+ are as follows:*

1) if A_1 is a block of type 2, then

1a) for indices $j = 1, 2, \ldots, n_1 - 1$ and $i = 1, 2, \ldots, j$:

$$z_{ij}^{(1)} = \frac{(-1)^{i+j} \sum\limits_{k=1}^{n_1-j} \left(\prod\limits_{s=j}^{n_1-k} \frac{1}{r_s^{(1)}} \right) \left(\prod\limits_{s=n_1-k+1}^{n_1-1} r_s^{(1)} \right)}{\prod\limits_{s=1}^{i-1} r_s^{(1)} \cdot d_j \sum\limits_{k=1}^{n_1} \left(\prod\limits_{s=1}^{n_1-k} \frac{1}{r_s^{(1)}} \right) \left(\prod\limits_{s=n_1-k+1}^{n_1-1} r_s^{(1)} \right)}; \tag{2.3.44}$$

1b) for indices $j = 1, 2, \ldots, n_1 - 1$ and $i = j + 1, j + 2, \ldots, n_1$:

$$z_{ij}^{(1)} = \frac{(-1)^{i+j+1} \left(\prod\limits_{s=i}^{n_1-1} r_s^{(1)} \right) \sum\limits_{k=1}^{j} \left(\prod\limits_{s=1}^{k-1} \frac{1}{r_s^{(1)}} \right) \left(\prod\limits_{s=k}^{j-1} r_s^{(1)} \right)}{d_j \sum\limits_{k=1}^{n_1} \left(\prod\limits_{s=1}^{n_1-k} \frac{1}{r_s^{(1)}} \right) \left(\prod\limits_{s=n_1-k+1}^{n_1-1} r_s^{(1)} \right)}; \tag{2.3.45}$$

1c) for index $j = n_1$:

$$z_{i\,n_1}^{(1)} = 0, \quad i = 1, 2, \ldots, n_1; \tag{2.3.46}$$

2) if A_1 is a block of type 3, then

2a) for $n_1 = 1$:

$$Z_1 = [0]_{1\times 1}; \tag{2.3.47}$$

2b) for $n_1 \geq 2$:

$$\begin{aligned} z_{i\,i-1}^{(1)} &= \frac{1}{b_{i-1}}, \quad i = 2, 3, \ldots, n_1; \\ z_{ij}^{(1)} &= 0, \ \textit{in other cases.} \end{aligned} \tag{2.3.48}$$

Below is an example to illustrate Lemma 2.3.8.

Example 2.3.1 Let the block A_1 be an $n_1 \times n_1$ bidiagonal matrix

$$A_1 = \begin{bmatrix} 1 & 1 & & \\ & \ddots & \ddots & \\ & & 1 & 1 \\ & & & 0 \end{bmatrix}.$$

The entries of the first $n_1 - 1$ columns of the block $Z_1 = A_1^+$ are as follows:

$$z_{ij}^{(1)} = \begin{cases} (-1)^{i+j} \left(1 - \dfrac{j}{n_1}\right), \ i = 1, 2, \ldots, j, \\[2ex] (-1)^{i+j+1} \dfrac{j}{n_1}, \ i = j+1, j+2, \ldots, n_1 \end{cases}, \quad j = 1, 2, \ldots, n_1 - 1. \ \diamond$$

Example 2.3.2 Consider the following 4×4 block

$$A_1 = \begin{bmatrix} 2 & 3 & & \\ & -3 & 1 & \\ & & 4 & 7 \\ & & & 0 \end{bmatrix}.$$

Computations by formulae (2.3.44) – (2.3.46) give

$$Z_1 = \begin{bmatrix} 0.4259 & 0.3930 & -0.0242 & 0 \\ 0.0494 & -0.2620 & 0.0161 & 0 \\ 0.1481 & 0.2140 & 0.0484 & 0 \\ -0.0846 & -0.1223 & 0.1152 & 0 \end{bmatrix}. \quad \diamond$$

2.4 Inversion of a pattern matrix $L(\varepsilon)$

Addressing the structure (2.2.13) of the matrices $L_k(\varepsilon)$, for $k \geq 2$, let us consider a pattern tridiagonal $m \times m$ matrix

$$L(\varepsilon) = \mathcal{A}^T\mathcal{A} + \mathcal{B}^T\mathcal{B} + \varepsilon I =$$

$$\begin{bmatrix} d_1^2+\Delta^2+\varepsilon & b_1d_1 & & & \\ b_1d_1 & d_2^2+b_1^2+\varepsilon & b_2d_2 & 0 & \\ & \ddots & \ddots & \ddots & \\ & 0 & b_{m-2}d_{m-2} & d_{m-1}^2+b_{m-2}^2+\varepsilon & b_{m-1}d_{m-1} \\ & & & b_{m-1}d_{m-1} & d_m^2+b_{m-1}^2+\varepsilon \end{bmatrix}, \tag{2.4.1}$$

which is defined by involvement of pattern matrices

$$\mathcal{A} = \begin{bmatrix} d_1 & b_1 & & & \\ & d_2 & b_2 & 0 & \\ & & \ddots & \ddots & \\ & 0 & & d_{m-1} & b_{m-1} \\ & & & & d_m \end{bmatrix} \tag{2.4.2}$$

of the size $m \times m$ and

$$\mathcal{B} = \begin{bmatrix} 0 & 0 & \dots & 0 \\ \vdots & \vdots & \dots & \vdots \\ 0 & 0 & \dots & 0 \\ \Delta & 0 & \dots & 0 \end{bmatrix} \tag{2.4.3}$$

of the size $l \times m$. We assume that the entries $b_1, b_2, \dots, b_{m-1}$ in the matrix $\mathcal{A}$ as well as the entry Δ in the matrix $\mathcal{B}$ are nonzero.

Remark 2.4.1 *In order to unify records of formulae, we set* $b_0 = \Delta$.

To derive formulae for the entries of inverse matrix $L(\varepsilon)^{-1}$, we will distinguish three cases associated with the arrangement of zeros on the main diagonal of the matrix $\mathcal{A}$. Namely, we will consider matrices $\mathcal{A}$ of the following types:

type 1: $m \geq 1$ and $d_1, d_2, \ldots, d_m \neq 0$;

type 2: $m \geq 2$ and $d_1, d_2, \ldots, d_{m-1} \neq 0,\ d_m = 0$;

type 3: $m \geq 1$ and $d_1 = d_2 = \ldots = d_m = 0$.

Emphasize that we give this classification in accordance with the types of diagonal blocks A_k in the block representation (2.1.3) of the initial matrix A.

• Type 1 (for $m = 1$).

As obviously follows from (2.4.1),

$$L(\varepsilon)^{-1} = \left[\frac{1}{d_1^2 + \Delta^2 + \varepsilon}\right]_{1\times1}. \tag{2.4.4}$$

• Types 1 and 2 (for $m \geq 2$).

The difference between types 1 and 2 just is in the value of the entry d_m. Therefore, it is conveniently to consider these cases together.

Having the matrix A from (2.4.2), let us introduce the following notation:

$$r_s \equiv \frac{b_s}{d_s}, \quad s = 1, 2, \ldots, m-1; \quad r_0 = r_m = 1. \tag{2.4.5}$$

To invert the matrix $L(\varepsilon)$, we apply again the computational procedure **3d/inv** from the Section 2.2. Comparing the records of the matrices $L(\varepsilon)$ and C given in (2.4.1) and (2.2.14), respectively, we have

$$c_{ii} = d_i^2 + b_{i-1}^2 + \varepsilon\,,\ i = 1, 2, \ldots, m \tag{2.4.6}$$

and

$$c_{i\,i+1} = b_i d_i\,,\ i = 1, 2, \ldots, m-1; \quad c_{i\,i-1} = b_{i-1} d_{i-1}\,,\ i = 2, 3, \ldots, m. \tag{2.4.7}$$

Let us figure out the dependence on ε of the quantities successively computed in the referred procedure **3d/inv**.

Consider first the quantities f_i, g_i and h_i introduced in (2.2.15). Using the expressions (2.4.6) and (2.4.7), we get

$$f_i = \overset{\circ}{f}_i + O(\varepsilon)\,,\ i = 2, 3, \ldots, m, \quad \text{where } \overset{\circ}{f}_i = \frac{d_i^2 + b_{i-1}^2}{b_{i-1} d_{i-1}}; \tag{2.4.8}$$

$$g_i = \frac{b_i d_i}{b_{i-1} d_{i-1}}\,,\ i = 2, 3, \ldots, m-1; \tag{2.4.9}$$

$$h_i = \overset{\circ}{h}_i + O(\varepsilon)\,,\ i = 1, 2, \ldots, m-1, \quad \text{where } \overset{\circ}{h}_i = \frac{d_i^2 + b_{i-1}^2}{b_i d_i}. \tag{2.4.10}$$

Next, consider the quantities μ_i and ν_i recursively computed in (2.2.16) and (2.2.17), respectively.

The following statement is readily obtained using the relations (2.2.16) and formulae (2.4.8), (2.4.9). Moreover, it can be considered as obvious consequence of Lemma 2.3.1 .

Lemma 2.4.1 *The quantities μ_i are represented as*

$$\mu_m = \overset{\circ}{\mu}_m ; \quad \mu_i = \overset{\circ}{\mu}_i + O(\varepsilon) , \ 1 \le i \le m-1 , \tag{2.4.11}$$

where the quantities $\overset{\circ}{\mu}_i$ satisfy the following recurrence relations:

$$\begin{aligned} &\overset{\circ}{\mu}_m = 1 , \ \overset{\circ}{\mu}_{m-1} = -\overset{\circ}{f}_m , \\ &\overset{\circ}{\mu}_i = -\overset{\circ}{f}_{i+1} \overset{\circ}{\mu}_{i+1} - g_{i+1} \overset{\circ}{\mu}_{i+2} , \ i = m-2, m-3, \ldots, 1. \end{aligned} \tag{2.4.12}$$

At the same time, we can get a closed form representation of the quantities $\overset{\circ}{\mu}_i$ computed by the recursion (2.4.12).

Lemma 2.4.2 *The quantities $\overset{\circ}{\mu}_i$ can be written in the form*

$$\overset{\circ}{\mu}_i = (-1)^{m-i} d_m^2 \sum_{k=i}^{m} \frac{1}{d_k^2} \left(\prod_{s=i}^{k-1} r_s \right) \left(\prod_{s=k}^{m-1} \frac{1}{r_s} \right) , \quad i = 1, 2, \ldots, m . \tag{2.4.13}$$

Proof. First, the value $\overset{\circ}{\mu}_m = 1$ conforms to the record (2.4.13). Then, according to (2.4.12) and (2.4.8),

$$\overset{\circ}{\mu}_{m-1} = -\overset{\circ}{f}_m = -\frac{d_m^2 + b_{m-1}^2}{b_{m-1} d_{m-1}} = -\frac{b_{m-1}}{d_{m-1}} - d_m^2 \frac{1}{b_{m-1} d_{m-1}} = -d_m^2 \left(\frac{r_{m-1}}{d_m^2} + \frac{1}{r_{m-1}} \frac{1}{d_{m-1}^2} \right) .$$

Further reasoning is carried out by induction. Using the expressions (2.4.8) and (2.4.9), proceeding from (2.4.12) we have

$$\begin{aligned} \overset{\circ}{\mu}_i &= -\overset{\circ}{f}_{i+1} \overset{\circ}{\mu}_{i+1} - g_{i+1} \overset{\circ}{\mu}_{i+2} \\ &= -\frac{d_{i+1}^2 + b_i^2}{b_i d_i} (-1)^{m-i-1} d_m^2 \sum_{k=i+1}^{m} \frac{1}{d_k^2} \left(\prod_{s=i+1}^{k-1} r_s \right) \left(\prod_{s=k}^{m-1} \frac{1}{r_s} \right) \\ &\quad - \frac{b_{i+1} d_{i+1}}{b_i d_i} (-1)^{m-i-2} d_m^2 \sum_{k=i+2}^{m} \frac{1}{d_k^2} \left(\prod_{s=i+2}^{k-1} r_s \right) \left(\prod_{s=k}^{m-1} \frac{1}{r_s} \right) \\ &= (-1)^{m-i} d_m^2 \frac{d_{i+1}^2 + b_i^2}{b_i d_i} \frac{1}{r_i} \sum_{k=i+1}^{m} \frac{1}{d_k^2} \left(\prod_{s=i}^{k-1} r_s \right) \left(\prod_{s=k}^{m-1} \frac{1}{r_s} \right) \\ &\quad - (-1)^{m-i} d_m^2 \frac{b_{i+1} d_{i+1}}{b_i d_i} \frac{1}{r_i} \frac{1}{r_{i+1}} \sum_{k=i+2}^{m} \frac{1}{d_k^2} \left(\prod_{s=i}^{k-1} r_s \right) \left(\prod_{s=k}^{m-1} \frac{1}{r_s} \right) . \end{aligned}$$

We continue to transform:

$$\begin{aligned}
\overset{\circ}{\mu}_i &= (-1)^{m-i}\, d_m^2 \left(1+\frac{d_{i+1}^2}{b_i^2}\right)\left[\sum_{k=i}^{m}\frac{1}{d_k^2}\left(\prod_{s=i}^{k-1} r_s\right)\left(\prod_{s=k}^{m-1}\frac{1}{r_s}\right)-\frac{1}{d_i^2}\prod_{s=i}^{m-1}\frac{1}{r_s}\right] \\
&\quad -(-1)^{m-i}\, d_m^2\,\frac{d_{i+1}^2}{b_i^2}\left[\sum_{k=i}^{m}\frac{1}{d_k^2}\left(\prod_{s=i}^{k-1} r_s\right)\left(\prod_{s=k}^{m-1}\frac{1}{r_s}\right)-\frac{1}{d_i^2}\left(1+\frac{b_i^2}{d_{i+1}^2}\right)\prod_{s=i}^{m-1}\frac{1}{r_s}\right] \\
&= (-1)^{m-i}\, d_m^2 \sum_{k=i}^{m}\frac{1}{d_k^2}\left(\prod_{s=i}^{k-1} r_s\right)\left(\prod_{s=k}^{m-1}\frac{1}{r_s}\right).
\end{aligned}$$

The Lemma has been proved. □

The next statement is a simple consequence of the formula (2.4.13).

Corollary 2.4.1 *The relation*

$$\overset{\circ}{\mu}_i = -r_i\,\overset{\circ}{\mu}_{i+1} + \frac{d_m^2}{d_i^2}\frac{1}{\alpha_i}, \quad i=1,2,\ldots,m-1, \tag{2.4.14}$$

where

$$\alpha_i \equiv (-1)^{m-i}\prod_{s=i}^{m-1} r_s, \quad i=1,2,\ldots,m-1, \tag{2.4.15}$$

holds.

Remark 2.4.2 *The quantities α_i defined in (2.4.15) can be computed recursively:*

$$\begin{aligned}
&\alpha_m = 1, \\
&\alpha_i = -r_i\alpha_{i+1}, \; i = m-1, m-2, \ldots, 1.
\end{aligned} \tag{2.4.16}$$

Now consider the quantities ν_i defined in (2.2.17). A statement similar to Lemma 2.4.1 holds true. It can be readily obtained using the relations (2.2.17) and formulae (2.4.9), (2.4.10). It can also be considered as obvious consequence of Lemma 2.3.3.

Lemma 2.4.3 *The quantities ν_i are represented as*

$$\nu_1 = \overset{\circ}{\nu}_1; \quad \nu_i = \overset{\circ}{\nu}_i + O(\varepsilon), \; 2 \le i \le m, \tag{2.4.17}$$

where the quantities $\overset{\circ}{\nu}_i$ satisfy the following recurrence relations:

$$\begin{aligned}
&\overset{\circ}{\nu}_1 = 1, \; \overset{\circ}{\nu}_2 = -\overset{\circ}{h}_1, \\
&\overset{\circ}{\nu}_i = -\overset{\circ}{h}_{i-1}\overset{\circ}{\nu}_{i-1} - \frac{1}{g_{i-1}}\overset{\circ}{\nu}_{i-2}, \quad i = 3,4,\ldots,m.
\end{aligned} \tag{2.4.18}$$

The quantities $\overset{\circ}{\nu}_i$ can be given in a closed form as well.

Lemma 2.4.4 *The quantities $\overset{\circ}{\nu}_i$ are written in the form*

$$\overset{\circ}{\nu}_i = (-1)^{i+1}\Delta^2 \sum_{k=0}^{i-1} \frac{1}{b_k^2} \left(\prod_{s=1}^{k} r_s\right) \left(\prod_{s=k+1}^{i-1} \frac{1}{r_s}\right), \quad i = 1, 2, \dots, m. \tag{2.4.19}$$

Proof. First of all remind that we have set $b_0 = \Delta$ (see Remark 2.4.1). Since $\overset{\circ}{\nu}_1 = 1$, this is consistent with formula (2.4.19). Then, according to (2.4.18) and (2.4.10),

$$\overset{\circ}{\nu}_2 = -\overset{\circ}{h}_1 = -\frac{d_1^2 + b_0^2}{b_1 d_1} = \frac{d_1}{b_1} - b_0^2 \frac{1}{b_1 d_1} = -\Delta^2 \left(\frac{1}{\Delta^2} \frac{1}{r_1} + \frac{1}{b_1^2} r_1\right).$$

Further reasoning is carried out by induction. Using the expressions (2.4.9) and (2.4.10), proceeding from the relation (2.4.18) we have

$$\begin{aligned}
\overset{\circ}{\nu}_i &= -\overset{\circ}{h}_{i-1} \overset{\circ}{\nu}_{i-1} - \frac{1}{g_{i-1}} \overset{\circ}{\nu}_{i-2} \\
&= -\frac{d_{i-1}^2 + b_{i-2}^2}{b_{i-1} d_{i-1}} (-1)^i \, \Delta^2 \sum_{k=0}^{i-2} \frac{1}{b_k^2} \left(\prod_{s=1}^{k} r_s\right) \left(\prod_{s=k+1}^{i-2} \frac{1}{r_s}\right) \\
&\quad -\frac{b_{i-2} d_{i-2}}{b_{i-1} d_{i-1}} (-1)^{i-1} \, \Delta^2 \sum_{k=0}^{i-3} \frac{1}{b_k^2} \left(\prod_{s=1}^{k} r_s\right) \left(\prod_{s=k+1}^{i-3} \frac{1}{r_s}\right) \\
&= (-1)^{i+1} \Delta^2 \frac{d_{i-1}^2 + b_{i-2}^2}{b_{i-1} d_{i-1}} r_{i-1} \sum_{k=0}^{i-2} \frac{1}{b_k^2} \left(\prod_{s=1}^{k} r_s\right) \left(\prod_{s=k+1}^{i-1} \frac{1}{r_s}\right) \\
&\quad -(-1)^{i+1} \Delta^2 \frac{b_{i-2} d_{i-2}}{b_{i-1} d_{i-1}} r_{i-2} r_{i-1} \sum_{k=0}^{i-3} \frac{1}{b_k^2} \left(\prod_{s=1}^{k} r_s\right) \left(\prod_{s=k+1}^{i-1} \frac{1}{r_s}\right).
\end{aligned}$$

Continuing transformations, we obtain

$$\begin{aligned}
\overset{\circ}{\nu}_i &= (-1)^{i+1} \Delta^2 \left(1 + \frac{b_{i-2}^2}{d_{i-1}^2}\right) \left[\sum_{k=0}^{i-1} \frac{1}{b_k^2} \left(\prod_{s=1}^{k} r_s\right) \left(\prod_{s=k+1}^{i-1} \frac{1}{r_s}\right) - \frac{1}{b_{i-1}^2} \prod_{s=1}^{i-1} r_s\right] \\
&\quad -(-1)^{i+1} \Delta^2 \frac{b_{i-2}^2}{d_{i-1}^2} \left[\sum_{k=0}^{i-1} \frac{1}{b_k^2} \left(\prod_{s=1}^{k} r_s\right) \left(\prod_{s=k+1}^{i-1} \frac{1}{r_s}\right) - \frac{1}{b_{i-1}^2} \left(1 + \frac{d_{i-1}^2}{b_{i-2}^2}\right) \prod_{s=1}^{i-1} r_s\right] \\
&= (-1)^{i+1} \Delta^2 \sum_{k=0}^{i-1} \frac{1}{b_k^2} \left(\prod_{s=1}^{k} r_s\right) \left(\prod_{s=k+1}^{i-1} \frac{1}{r_s}\right).
\end{aligned}$$

The Lemma has been proved. □

The next statement is a simple consequence of the formula (2.4.19).

Corollary 2.4.2 *The relation*

$$\overset{\circ}{\nu}_{i+1} = -\frac{1}{r_i}\overset{\circ}{\nu}_i + \frac{\Delta^2}{b_i^2}\beta_i\,, \quad i = 1, 2, \ldots, m-1\,, \tag{2.4.20}$$

where

$$\beta_i \equiv (-1)^i \prod_{s=1}^{i} r_s\,, \quad i = 1, 2, \ldots, m-1\,, \tag{2.4.21}$$

holds.

Remark 2.4.3 *The quantities α_i defined in (2.4.21) can be computed recursively:*

$$\begin{aligned} \beta_1 &= -r_1, \\ \beta_{i+1} &= -r_{i+1}\beta_i\,,\; i = 1, 2, \ldots, m-2\,. \end{aligned} \tag{2.4.22}$$

Further, let us turn now to the quantity t defined in (2.2.18). Since $c_{11} = d_1^2 + \Delta^2 + \varepsilon$, $c_{12} = b_1 d_1$ (see (2.4.6) and (2.4.7)), then having the representation (2.4.11) of the quantity $\overset{\circ}{\mu}_i$ we get

$$\begin{aligned} t &= (c_{11}\mu_1 + c_{12}\mu_2)^{-1} = ((d_1^2 + \Delta^2 + \varepsilon)(\overset{\circ}{\mu}_1 + O(\varepsilon)) + b_1 d_1(\overset{\circ}{\mu}_2 + O(\varepsilon)))^{-1} \\ &= (d_1^2(\overset{\circ}{\mu}_1 + r_1\overset{\circ}{\mu}_2) + \Delta^2\overset{\circ}{\mu}_1 + O(\varepsilon))^{-1} = (q + O(\varepsilon))^{-1}, \end{aligned}$$

where

$$q \equiv d_1^2(\overset{\circ}{\mu}_1 + r_1\overset{\circ}{\mu}_2) + \Delta^2\overset{\circ}{\mu}_1\,. \tag{2.4.23}$$

As follows from the relation (2.4.14),

$$\overset{\circ}{\mu}_1 + r_1\overset{\circ}{\mu}_2 = \frac{d_m^2}{d_1^2}\frac{1}{\alpha_1}$$

and, consequently,

$$q = \frac{d_m^2}{\alpha_1} + \Delta^2\overset{\circ}{\mu}_1\,. \tag{2.4.24}$$

If we substitute into the right hand side of the previous equality the expressions (2.4.13) and (2.4.15) of the quantities $\overset{\circ}{\mu}_1$ and α_1, respectively, then we arrive at the following assertion.

Lemma 2.4.5 *The quantity t is written as*

$$t = (q + O(\varepsilon))^{-1}\,, \tag{2.4.25}$$

where

$$q = (-1)^{m-1} d_m^2 \left[\prod_{s=1}^{m-1}\frac{1}{r_s} + \Delta^2\sum_{k=1}^{m}\frac{1}{d_k^2}\left(\prod_{s=1}^{k-1} r_s\right)\left(\prod_{s=k}^{m-1}\frac{1}{r_s}\right)\right]. \tag{2.4.26}$$

Finally, the entries of the inverse matrix

$$L(\varepsilon)^{-1} = [x_{ij}]_{m\times m} \tag{2.4.27}$$

are computed, by the formulae (2.2.19) and (2.2.20). First we find the entries of the lower triangular part of the matrix, including the main diagonal:

$$x_{ij} = \mu_i \nu_j t\,,\ i = j, j+1, \ldots, m;\ j = 1, 2, \ldots, m\,. \tag{2.4.28}$$

Then we get the entries of the upper triangular part:

$$x_{ij} = \mu_j \nu_i t\,,\ i = 1, 2, \ldots, j-1;\ j = 2, 3, \ldots, m\,. \tag{2.4.29}$$

At the same time we note that due to the symmetry of the matrix $L(\varepsilon)^{-1}$ practically it is not necessary to perform computations by the formula (2.4.29), we can simply set $x_{ij} = x_{ji}$.

- Type 3.

As follows from (2.4.1), the matrix $L(\varepsilon)$ in this case is diagonal:

$$L(\varepsilon) = \begin{bmatrix} \Delta^2+\varepsilon & & & & \\ & b_1^2+\varepsilon & & 0 & \\ & & \ddots & & \\ & 0 & & b_{m-2}^2+\varepsilon & \\ & & & & b_{m-1}^2+\varepsilon \end{bmatrix}.$$

Hence,

$$L(\varepsilon)^{-1} = \begin{bmatrix} \dfrac{1}{\Delta^2+\varepsilon} & & & & \\ & \dfrac{1}{b_1^2+\varepsilon} & & 0 & \\ & & \ddots & & \\ & 0 & & \dfrac{1}{b_{m-2}^2+\varepsilon} & \\ & & & & \dfrac{1}{b_{m-1}^2+\varepsilon} \end{bmatrix}. \tag{2.4.30}$$

Thus, the inversion process of the matrix $L(\varepsilon)$ is fully described.

A plan of our subsequent steps consists in the following. The main objective is to derive formulae for the entries of the blocks Z_k and H_k involved in the block representation (2.2.5) of the matrix A^+. These blocks are defined by means of the equalities (2.2.6) and (2.2.7) respectively. The block Z_1 has been computed in the Section 2.3. To get the solution of the problem for the values $k \geq 2$, we consider more general problem of computing the pattern matrices

$$Z = \lim_{\varepsilon\to+0} L(\varepsilon)^{-1} \mathcal{A}^T \tag{2.4.31}$$

and

$$H = \lim_{\varepsilon \to +0} L(\varepsilon)^{-1} \mathcal{B}^T \tag{2.4.32}$$

where the matrices $L(\varepsilon)$, $\mathcal{A}$, $\mathcal{B}$ are given in (2.4.1) – (2.4.2).

Let

$$Z = [z_{ij}]_{m \times m}\,, \qquad H = [h_{ij}]_{m \times l}\,.$$

Further study we will carry out in accordance with the types1, 2 and 3 specified in the Section 2.4.

2.5 Calculation of the pattern matrices Z and H: type 1

If $m = 1$, then the matrices Z and H have a simple view:

$$Z = \left[\frac{d_1}{d_1^2 + \Delta^2}\right]_{1\times 1}, \qquad H = \left[0 \ldots 0 \frac{\Delta}{d_1^2 + \Delta^2}\right]_{1\times l}; \tag{2.5.1}$$

notice, that $H = \left[\dfrac{\Delta}{d_1^2 + \Delta^2}\right]_{1\times 1}$ when $l = 1$. It can be easily obtained from (2.4.2) – (2.4.4).

The case $m \geq 2$ is more complicated. Let us start with computation of the matrix Z. For this we introduce a matrix

$$L(\varepsilon)^{-1} \mathcal{A}^T \equiv Y(\varepsilon) = [y_{ij}(\varepsilon)]_{m \times m}\,. \tag{2.5.2}$$

Based on (2.4.31), we have

$$z_{ij} = \lim_{\varepsilon \to +0} y_{ij}(\varepsilon)\,, \quad i, j = 1, 2, \ldots, m. \tag{2.5.3}$$

As it follows from (2.5.2), for indices $1 \leq j \leq m-1$ the entries $y_{ij}(\varepsilon)$ of the matrix $Y(\varepsilon)$ are calculated by the rule

$$y_{ij}(\varepsilon) = x_{ij} d_j + x_{i\,j+1} b_j, \quad i = 1, 2, \ldots, m \tag{2.5.4}$$

(the value $j = m$ of the index will be considered separately). For a fixed index j in the range $1 \leq j \leq m-1$ we will distinguish two cases: $i = 1, 2, \ldots, j$ and $i = j+1, j+2, \ldots, m$.

- Indices $i = 1, 2, \ldots, j$.

Taking advantage of the expression (2.4.29) for entries x_{ij} of the matrix $L(\varepsilon)^{-1}$, from (2.5.4) we write

$$y_{ij}(\varepsilon) = t\nu_i(\mu_j d_j + \mu_{j+1} b_j). \tag{2.5.5}$$

Then, using the representations (2.4.11) of the quantities $\mathring{\mu}_i$, we have

$$\mu_j d_j + \mu_{j+1} b_j = (\mathring{\mu}_j\, d_j + \mathring{\mu}_{j+1}\, b_j) + O(\varepsilon).$$

Substituting the last expression as well as the representations (2.4.17) and (2.4.25) of the quantities ν_i and t, respectively, into the right hand side of the equality (2.5.5) yields

$$y_{ij}(\varepsilon) = \frac{(\overset{\circ}{\nu}_i + O(\varepsilon))((\overset{\circ}{\mu}_j d_j + \overset{\circ}{\mu}_{j+1} b_j) + O(\varepsilon))}{q + O(\varepsilon)}.$$

From here, by taking limit as $\varepsilon \to +0$, according to (2.5.3) we obtain

$$z_{ij} = \frac{\overset{\circ}{\nu}_i (\overset{\circ}{\mu}_j d_j + \overset{\circ}{\mu}_{j+1} b_j)}{q}, \quad i = 1, 2, \dots, j.$$

As follows from the relation (2.4.14),

$$\overset{\circ}{\mu}_j d_j + \overset{\circ}{\mu}_{j+1} b_j = d_j(\overset{\circ}{\mu}_j + r_j \overset{\circ}{\mu}_{j+1}) = \frac{d_m^2}{d_j}\frac{1}{\alpha_j}.$$

Therefore,

$$z_{ij} = \frac{\overset{\circ}{\nu}_i d_m^2}{q}\frac{1}{d_j}\frac{1}{\alpha_j}, \quad i = 1, 2, \dots, j. \tag{2.5.6}$$

To get the final closed form expression for the entries z_{ij}, let us substitute into the equality (2.5.6) the expressions (2.4.15), (2.4.19) and (2.4.26) of the quantities α_j, $\overset{\circ}{\nu}_i$ and q, respectively. As a result, for the indices $1 \le j \le m-1$ and $i = 1, 2, \dots, j$ we find

$$z_{ij} = \frac{(-1)^{i+j}\Delta^2 \sum\limits_{k=0}^{i-1} \frac{1}{b_k^2}\left(\prod\limits_{s=1}^{k} r_s\right)\left(\prod\limits_{s=k+1}^{i-1} \frac{1}{r_s}\right)}{d_j\left(\prod\limits_{s=j}^{m-1} r_s\right)\left[\prod\limits_{s=1}^{m-1}\frac{1}{r_s} + \Delta^2 \sum\limits_{k=1}^{m}\frac{1}{d_k^2}\left(\prod\limits_{s=1}^{k-1} r_s\right)\left(\prod\limits_{s=k}^{m-1}\frac{1}{r_s}\right)\right]}. \tag{2.5.7}$$

- Indices $i = j+1, j+2, \dots, m$.

By making use of the expression (2.4.27) for the entries of the matrix $L(\varepsilon)^{-1}$, from (2.5.4) it follows that

$$y_{ij}(\varepsilon) = t\mu_i(\nu_j d_j + \nu_{j+1} b_j). \tag{2.5.8}$$

Having the representations (2.4.17) of the quantities ν_i, we get

$$\nu_j d_j + \nu_{j+1} b_j = (\overset{\circ}{\nu}_j d_j + \overset{\circ}{\nu}_{j+1} b_j) + O(\varepsilon).$$

Then, substituting the previous expression as well as the representations (2.4.11) and (2.4.25) of the quantities μ_i and t, respectively, into the right hand side of the equality (2.5.8) yields

$$y_{ij}(\varepsilon) = \frac{(\overset{\circ}{\mu}_i + O(\varepsilon))((\overset{\circ}{\nu}_j d_j + \overset{\circ}{\nu}_{j+1} b_j) + O(\varepsilon))}{q + O(\varepsilon)}.$$

From here, by taking limit as $\varepsilon \to +0$, according to (2.5.3) we obtain

$$z_{ij} = \frac{\overset{\circ}{\mu}_i\,(\overset{\circ}{\nu}_j\, d_j + \overset{\circ}{\nu}_{j+1}\, b_j)}{q}, \qquad i = j+1, j+2, \ldots, m\,.$$

By the relation (2.4.20),

$$\overset{\circ}{\nu}_j\, d_j + \overset{\circ}{\nu}_{j+1}\, b_j = b_j \left(\frac{1}{r_j}\, \overset{\circ}{\nu}_j + \overset{\circ}{\nu}_{j+1} \right) = \frac{\Delta^2}{b_j} \beta_j\,.$$

Thus,

$$z_{ij} = \frac{\overset{\circ}{\mu}_i\, \Delta^2}{q\ \ b_j} \beta_j\,, \qquad i = j+1, j+2, \ldots, m\,. \tag{2.5.9}$$

To get the final closed form expression for the entries z_{ij}, let us substitute into the equality (2.5.9) the expressions (2.4.13), (2.4.21) and (2.4.26) of the quantities $\overset{\circ}{\mu}_i$, β_j and q, respectively. As a result, for the indices $1 \le j \le m-1$ and $i = j+1, j+2, \ldots, m$ we find

$$z_{ij} = \frac{(-1)^{i+j+1}\Delta^2 \left(\prod\limits_{s=1}^{j} r_s \right) \sum\limits_{k=i}^{m} \frac{1}{d_k^2} \left(\prod\limits_{s=i}^{k-1} r_s \right) \left(\prod\limits_{s=k}^{m-1} \frac{1}{r_s} \right)}{b_j \left[\prod\limits_{s=1}^{m-1} \frac{1}{r_s} + \Delta^2 \sum\limits_{k=1}^{m} \frac{1}{d_k^2} \left(\prod\limits_{s=1}^{k-1} r_s \right) \left(\prod\limits_{s=k}^{m-1} \frac{1}{r_s} \right) \right]}\,. \tag{2.5.10}$$

It remains to derive formula for the entries of the last column of the matrix Z. As follows from (2.5.2), the entries $y_{im}(\varepsilon)$ of the matrix $Y(\varepsilon)$ are

$$y_{im}(\varepsilon) = x_{im} d_m\,, \qquad i = 1, 2, \ldots, m\,.$$

According to the equality (2.4.28), $x_{im} = \mu_m \nu_i t$, and since $\mu_m = 1$ (see (2.2.16)), then

$$y_{im}(\varepsilon) = \nu_i t d_m\,, \qquad i = 1, 2, \ldots, m\,.$$

Further, using the representations (2.4.17) and (2.4.25) of the quantities ν_i and t, respectively, we can write that

$$y_{im}(\varepsilon) = \frac{d_m(\overset{\circ}{\nu}_i + O(\varepsilon))}{q + O(\varepsilon)}, \qquad i = 1, 2, \ldots, m\,.$$

By taking limit in this equality, according to (2.5.3) we obtain

$$z_{im} = \frac{d_m\, \overset{\circ}{\nu}_i}{q}, \qquad i = 1, 2, \ldots, m\,. \tag{2.5.11}$$

Finally, substituting the expressions (2.4.19) and (2.5.26), for the indices $i = 1, 2, \ldots, m$ we find

$$z_{im} = \frac{(-1)^{m-i}\Delta^2 \sum\limits_{k=0}^{i-1} \frac{1}{b_k^2} \left(\prod\limits_{s=1}^{k} r_s \right) \left(\prod\limits_{s=k+1}^{i-1} \frac{1}{r_s} \right)}{d_m \left[\prod\limits_{s=1}^{m-1} \frac{1}{r_s} + \Delta^2 \sum\limits_{k=1}^{m} \frac{1}{d_k^2} \left(\prod\limits_{s=1}^{k-1} r_s \right) \left(\prod\limits_{s=k}^{m-1} \frac{1}{r_s} \right) \right]}\,. \tag{2.5.12}$$

Thus, in (2.5.1), (2.5.7), (2.5.10) and (2.5.12) we give the closed form expressions for the entries of the matrix Z.

Remark 2.5.1 *It is easy to see that the expression (2.5.1) for the entry of the matrix Z, when $m = 1$, follows from the more general formula (2.5.12). Further, the immediate verification shows that the formula (2.5.12) can be "incorporated" in the formula (2.5.7), if we extend the last one to the case $j = m$.*

Summarizing the above considerations, we arrive at the following statement.

Lemma 2.5.1 *The entries of the matrix $Z = [z_{ij}]_{m\times m}$ defined in (2.4.31) are as follows:*

1) for indices $j = 1, 2, \ldots, m$ and $i = 1, 2, \ldots, j$:

$$z_{ij} = \frac{(-1)^{i+j}\Delta^2 \sum_{k=0}^{i-1} \frac{1}{b_k^2}\left(\prod_{s=1}^{k} r_s\right)\left(\prod_{s=k+1}^{i-1}\frac{1}{r_s}\right)}{d_j\left(\prod_{s=j}^{m-1} r_s\right)\left[\prod_{s=1}^{m-1}\frac{1}{r_s} + \Delta^2\sum_{k=1}^{m}\frac{1}{d_k^2}\left(\prod_{s=1}^{k-1} r_s\right)\left(\prod_{s=k}^{m-1}\frac{1}{r_s}\right)\right]}; \tag{2.5.13}$$

2) for indices $j = 1, 2, \ldots, m-1$ and $i = j+1, j+2, \ldots, m$:

$$z_{ij} = \frac{(-1)^{i+j+1}\Delta^2\left(\prod_{s=1}^{j} r_s\right)\sum_{k=i}^{m}\frac{1}{d_k^2}\left(\prod_{s=i}^{k-1} r_s\right)\left(\prod_{s=k}^{m-1}\frac{1}{r_s}\right)}{b_j\left[\prod_{s=1}^{m-1}\frac{1}{r_s} + \Delta^2\sum_{k=1}^{m}\frac{1}{d_k^2}\left(\prod_{s=1}^{k-1} r_s\right)\left(\prod_{s=k}^{m-1}\frac{1}{r_s}\right)\right]}. \tag{2.5.14}$$

Let us turn now to the computation of the matrix H defined in (2.4.32). For this we introduce a matrix

$$L(\varepsilon)^{-1}\mathcal{B}^T \equiv W(\varepsilon) = [w_{ij}(\varepsilon)]_{n\times l}. \tag{2.5.15}$$

By the definition of the matrix H we can write that

$$h_{ij} = \lim_{\varepsilon\to+0} w_{ij}(\varepsilon), \quad i = 1, 2, \ldots, m,\ j = 1, 2, \ldots, l. \tag{2.5.16}$$

For $l \geq 2$, it can be easily seen that the first $l-1$ columns of the matrix $W(\varepsilon)$ are zero. This implies that the corresponding columns of the matrix H are also zeros, i.e. the matrix H has the following structure:

$$H = \begin{bmatrix} 0 & 0 & \ldots & 0 & h_{1l} \\ 0 & 0 & \ldots & 0 & h_{2l} \\ \vdots & \vdots & \ddots & \vdots & \vdots \\ 0 & 0 & \ldots & 0 & h_{ml} \end{bmatrix}. \tag{2.5.17}$$

The entries of the last column of the matrix $W(\varepsilon)$ are

$$w_{il}(\varepsilon) = x_{i1}\Delta, \quad i = 1, 2, \ldots, m.$$

Since $x_{i1} = \mu_i \nu_1 t$ (see (2.4.28)) and $\nu_1 = 1$ (see (2.2.17)), then

$$w_{il}(\varepsilon) = \mu_i t \Delta, \quad i = 1, 2, \ldots, m.$$

Further, using the representations (2.4.11) and (2.4.25) of the quantities μ_i and t, respectively, we have

$$w_{il}(\varepsilon) = \frac{\overset{\circ}{\mu}_i + O(\varepsilon)}{q + O(\varepsilon)} \Delta, \quad i = 1, 2, \ldots, m.$$

By taking limit in this equality, as $\varepsilon \to +0$, according to (2.5.16) we get that

$$h_{il} = \frac{\overset{\circ}{\mu}_i \Delta}{q}, \quad i = 1, 2, \ldots, m. \tag{2.5.18}$$

Substituting the expressions (2.4.13) and (2.4.26) of the quantities $\overset{\circ}{\mu}_i$ and q, respectively, into the right hand side of the previous equality, for the indices $i = 1, 2, \ldots, m$ we find

$$h_{il} = \frac{(-1)^{i+1} \Delta \sum\limits_{k=i}^{m} \frac{1}{d_k^2} \left(\prod\limits_{s=i}^{k-1} r_s \right) \left(\prod\limits_{s=k}^{m-1} \frac{1}{r_s} \right)}{\prod\limits_{s=1}^{m-1} \frac{1}{r_s} + \Delta^2 \sum\limits_{k=1}^{m} \frac{1}{d_k^2} \left(\prod\limits_{s=1}^{k-1} r_s \right) \left(\prod\limits_{s=k}^{m-1} \frac{1}{r_s} \right)}. \tag{2.5.19}$$

Thus, in (2.5.1), (2.5.17) and (2.5.19) we give the closed form expressions for the entries of the matrix H.

Remark 2.5.2 *It is easy to note that the expression for the entry h_{1l} of the matrix H, when $m = 1$ (see (2.5.1)), follows from the more general formula (2.5.19).*

Remark 2.5.3 *If $l = 1$, then the matrix H from (2.5.17) has the following view:*

$$H = \begin{bmatrix} h_{11} \\ h_{21} \\ \vdots \\ h_{m1} \end{bmatrix}.$$

Summarizing the above considerations, we arrive at the next statement.

Lemma 2.5.2 *The entries of the matrix $H = [h_{ij}]_{m \times l}$ defined in (2.4.32) are as follows:*

1) for indices $j = 1, 2, \ldots, l-1$ and $i = 1, 2, \ldots, m$:

$$h_{ij} = 0; \tag{2.5.20}$$

2) for indices $i = 1, 2, \ldots, m$:

$$h_{il} = \frac{(-1)^{i+1} \Delta \sum\limits_{k=i}^{m} \frac{1}{d_k^2} \left(\prod\limits_{s=i}^{k-1} r_s \right) \left(\prod\limits_{s=k}^{m-1} \frac{1}{r_s} \right)}{\prod\limits_{s=1}^{m-1} \frac{1}{r_s} + \Delta^2 \sum\limits_{k=1}^{m} \frac{1}{d_k^2} \left(\prod\limits_{s=1}^{k-1} r_s \right) \left(\prod\limits_{s=k}^{m-1} \frac{1}{r_s} \right)}. \tag{2.5.21}$$

Next example illustrates Lemmas 2.5.1 and 2.5.2.

Example 2.5.1 Consider a bidiagonal matrix which is partitioned into blocks as follows:

$$A = \left[\begin{array}{cccc|cccc} 2 & 3 & & & & & & \\ & -3 & 1 & & & & & \\ & & 4 & 7 & & & & \\ & & & 0 & 8 & & & \\ \hline & & & & 11 & 3 & & \\ & & & & & 9 & -5 & \\ & & & & & & 6 & 7 \\ & & & & & & & 3 \end{array}\right].$$

Taking advantage of MATLAB software system, we get

$$A^+ = \left[\begin{array}{cccc|cccc} 0.4259 & 0.3930 & -0.0242 & 0 & 0 & 0 & 0 & 0 \\ 0.0494 & -0.2620 & 0.0161 & 0 & 0 & 0 & 0 & 0 \\ 0.1481 & 0.2140 & 0.0484 & 0 & 0 & 0 & 0 & 0 \\ -0.0846 & -0.1223 & 0.1152 & 0 & 0 & 0 & 0 & 0 \\ \hline 0 & 0 & 0 & 0.0575 & 0.0491 & -0.0164 & -0.0136 & 0.0318 \\ 0 & 0 & 0 & -0.0797 & 0.0580 & 0.0918 & 0.0765 & -0.1785 \\ 0 & 0 & 0 & -0.1172 & 0.0853 & -0.0284 & 0.1430 & -0.3336 \\ 0 & 0 & 0 & 0.0849 & -0.0617 & 0.0206 & 0.0172 & 0.2933 \end{array}\right].$$

Let us take as matrices $\mathcal{A}$ and $\mathcal{B}$ right lower and right upper blocks of the matrix A, respectively, i.e.

$$\mathcal{A} = \begin{bmatrix} 11 & 3 & & \\ & 9 & -5 & \\ & & 6 & 7 \\ & & & 3 \end{bmatrix}, \quad \mathcal{B} = \begin{bmatrix} 0 & 0 & 0 & 0 \\ 0 & 0 & 0 & 0 \\ 0 & 0 & 0 & 0 \\ 8 & 0 & 0 & 0 \end{bmatrix}$$

(see (2.4.2) and (2.4.3)). Calculations by formulae (2.5.13), (2.5.14) and (2.5.20), (2.5.21) give us

$$Z = \begin{bmatrix} 0.0491 & -0.0164 & -0.0136 & 0.0318 \\ 0.0580 & 0.0918 & 0.0765 & -0.1785 \\ 0.0853 & -0.0284 & 0.1430 & -0.3336 \\ -0.0617 & 0.0206 & 0.0172 & 0.2933 \end{bmatrix}, \quad H = \begin{bmatrix} 0 & 0 & 0 & 0.0575 \\ 0 & 0 & 0 & -0.0797 \\ 0 & 0 & 0 & -0.1172 \\ 0 & 0 & 0 & 0.0849 \end{bmatrix}.$$

It is easy to see that the matrices Z and H coincide with right lower and left lower blocks of the matrix A^+, respectively. $\diamond$

2.6 Calculation of the pattern matrices Z and H: type 2

The difference between types 1 and 2 consists only in the value of diagonal entry d_m of the matrix $\mathcal{A}$ from (2.4.2). In the case of type 1 we assumed that $d_m \neq 0$ while in the type 2 we have $d_m = 0$. Therefore we can use formulae and expressions obtained in the previous section, substituting in them $d_m = 0$.

Let us start with the formula (2.5.13) which specifies the entries of the upper triangular part of the matrix Z. After simple transformations we obtain

$$z_{ij} = \frac{(-1)^{i+j}\Delta^2 \sum\limits_{k=0}^{i-1} \frac{1}{b_k^2} \left(\prod\limits_{s=1}^{k} r_s\right)\left(\prod\limits_{s=k+1}^{i-1} \frac{1}{r_s}\right)}{d_j \left(\prod\limits_{s=j}^{m-1} r_s\right)\left[\prod\limits_{s=1}^{m-1}\frac{1}{r_s} + \Delta^2 \sum\limits_{k=1}^{m-1}\frac{1}{d_k^2}\left(\prod\limits_{s=1}^{k-1} r_s\right)\left(\prod\limits_{s=k}^{m-1}\frac{1}{r_s}\right) + \frac{\Delta^2}{d_m^2}\prod\limits_{s=1}^{m-1} r_s\right]}$$

$$= \frac{(-1)^{i+j} d_m^2 \Delta^2 \sum\limits_{k=0}^{i-1} \frac{1}{b_k^2} \left(\prod\limits_{s=1}^{k} r_s\right)\left(\prod\limits_{s=k+1}^{i-1} \frac{1}{r_s}\right)}{d_j \left(\prod\limits_{s=j}^{m-1} r_s\right)\left[d_m^2\prod\limits_{s=1}^{m-1}\frac{1}{r_s} + d_m^2\Delta^2 \sum\limits_{k=1}^{m-1}\frac{1}{d_k^2}\left(\prod\limits_{s=1}^{k-1} r_s\right)\left(\prod\limits_{s=k}^{m-1}\frac{1}{r_s}\right) + \Delta^2\prod\limits_{s=1}^{m-1} r_s\right]}.$$

Substituting $d_m = 0$, we find that

$$z_{ij} = 0\,, \quad j = 1, 2, \ldots, m\,,\ i = 1, 2, \ldots, j\,.$$

Consider now the formula (2.5.14). The entry z_{ij} can be written as follows:

$$z_{ij} = \frac{(-1)^{i+j+1}\Delta^2 \left(\prod\limits_{s=1}^{j} r_s\right)\left[\sum\limits_{k=i}^{m-1}\frac{1}{d_k^2}\left(\prod\limits_{s=i}^{k-1} r_s\right)\left(\prod\limits_{s=k}^{m-1}\frac{1}{r_s}\right) + \frac{1}{d_m^2}\prod\limits_{s=i}^{m-1} r_s\right]}{b_j\left[\prod\limits_{s=1}^{m-1}\frac{1}{r_s} + \Delta^2\sum\limits_{k=1}^{m-1}\frac{1}{d_k^2}\left(\prod\limits_{s=1}^{k-1} r_s\right)\left(\prod\limits_{s=k}^{m-1}\frac{1}{r_s}\right) + \frac{\Delta^2}{d_m^2}\prod\limits_{s=1}^{m-1} r_s\right]}$$

$$= \frac{(-1)^{i+j+1}\Delta^2 \left(\prod\limits_{s=1}^{j} r_s\right)\left[d_m^2\sum\limits_{k=i}^{m-1}\frac{1}{d_k^2}\left(\prod\limits_{s=i}^{k-1} r_s\right)\left(\prod\limits_{s=k}^{m-1}\frac{1}{r_s}\right) + \prod\limits_{s=i}^{m-1} r_s\right]}{b_j\left[d_m^2\prod\limits_{s=1}^{m-1}\frac{1}{r_s} + d_m^2\Delta^2\sum\limits_{k=1}^{m-1}\frac{1}{d_k^2}\left(\prod\limits_{s=1}^{k-1} r_s\right)\left(\prod\limits_{s=k}^{m-1}\frac{1}{r_s}\right) + \Delta^2\prod\limits_{s=1}^{m-1} r_s\right]}.$$

Substituting $d_m = 0$ we get

$$z_{ij} = \frac{(-1)^{i+j+1}}{d_j}\prod_{s=j}^{i-1}\frac{1}{r_s}, \quad j = 1, 2, \ldots, m-1\,,\ i = j+1, j+2, \ldots, m\,.$$

Thus, we arrive at the following statement.

Lemma 2.6.1 *The entries of the matrix $Z = [z_{ij}]_{m\times m}$ defined in (2.4.31) are as follows:*

1) for indices $j = 1, 2, \ldots, m$ and $i = 1, 2, \ldots, j$:

$$z_{ij} = 0\,; \tag{2.6.1}$$

2) for indices $j = 1, 2, \ldots, m-1$ and $i = j+1, j+2, \ldots, m$:

$$z_{ij} = \frac{(-1)^{i+j+1}}{d_j} \prod_{s=j}^{i-1} \frac{1}{r_s}. \tag{2.6.2}$$

Observe that the process of calculating the lower triangular part of the matrix Z can be organized as follows. Consider fixed value of the index j from the range $1 \le j \le m-1$. For $i = j+1$, by (2.6.2) we obtain

$$z_{j+1\,j} = \frac{1}{d_j r_j} = \frac{1}{b_j}. \tag{2.6.3}$$

Further, for the subsequent values $i = j+2, j+3, \ldots, m$, again due to (2.6.2), the following relation holds:

$$z_{ij} = -\frac{z_{i-1\,j}}{r_{i-1}}. \tag{2.6.4}$$

Now consider the matrix H, the entries of which are specified in Lemma 2.5.2. Having the formula (2.5.21), as a result of simple transformations we have

$$h_{il} = \frac{(-1)^{i+1}\Delta\left[\sum\limits_{k=i}^{m-1}\frac{1}{d_k^2}\left(\prod\limits_{s=i}^{k-1} r_s\right)\left(\prod\limits_{s=k}^{m-1}\frac{1}{r_s}\right) + \frac{1}{d_m^2}\prod\limits_{s=i}^{m-1} r_s\right]}{\prod\limits_{s=1}^{m-1}\frac{1}{r_s} + \Delta^2\left[\sum\limits_{k=1}^{m-1}\frac{1}{d_k^2}\left(\prod\limits_{s=1}^{k-1} r_s\right)\left(\prod\limits_{s=k}^{m-1}\frac{1}{r_s}\right) + \frac{1}{d_m^2}\prod\limits_{s=1}^{m-1} r_s\right]}$$

$$= \frac{(-1)^{i+1}\Delta\left[d_m^2\sum\limits_{k=i}^{m-1}\frac{1}{d_k^2}\left(\prod\limits_{s=i}^{k-1} r_s\right)\left(\prod\limits_{s=k}^{m-1}\frac{1}{r_s}\right) + \prod\limits_{s=i}^{m-1} r_s\right]}{d_m^2\prod\limits_{s=1}^{m-1}\frac{1}{r_s} + \Delta^2\left[d_m^2\sum\limits_{k=1}^{m-1}\frac{1}{d_k^2}\left(\prod\limits_{s=1}^{k-1} r_s\right)\left(\prod\limits_{s=k}^{m-1}\frac{1}{r_s}\right) + \prod\limits_{s=1}^{m-1} r_s\right]}.$$

Substituting $d_m = 0$, we find

$$h_{il} = \frac{(-1)^{i+1}}{\Delta}\prod_{s=1}^{i-1}\frac{1}{r_s}, \qquad i = 1, 2, \ldots, m.$$

Summarizing the above considerations, we arrive at the following result.

Lemma 2.6.2 *The entries of the matrix $H = [h_{ij}]_{m\times l}$ defined in (2.4.32) are as follows:*

1) for indices $j = 1, 2, \ldots, l-1$ and $i = 1, 2, \ldots, m$:

$$h_{ij} = 0; \tag{2.6.5}$$

2) for indices $i = 1, 2, \ldots, m$:

$$h_{il} = \frac{(-1)^{i+1}}{\Delta}\prod_{s=1}^{i-1}\frac{1}{r_s}. \tag{2.6.6}$$

Referring to Lemma 2.6.2, should take into account the Remark 2.5.3.

The calculation of the last column of the matrix H can also be carried out by a recurrence relation. For $i = 1$, as follows from (2.6.6),

$$h_{1l} = \frac{1}{\Delta}. \tag{2.6.7}$$

Then, for the values $i = 2, 3, \dots, m$ we have

$$h_{il} = -\frac{h_{i-1\,l}}{r_{i-1}}. \tag{2.6.8}$$

Below we give an example to illustrate Lemmas 2.6.1 and 2.6.2.

Example 2.6.1 Consider a bidiagonal matrix which is partitioned into blocks as follows:

$$A = \left[\begin{array}{cccc|cccc} 2 & 3 & & & & & & \\ & -3 & 1 & & & & & \\ & & 4 & 7 & & & & \\ & & & 0 & 8 & & & \\ \hline & & & & 11 & 3 & & \\ & & & & & 9 & -5 & \\ & & & & & & 6 & 7 \\ & & & & & & & 0 \end{array}\right].$$

Taking advantage of MATLAB software system, we get

$$A^+ = \left[\begin{array}{cccc|cccc} 0.4259 & 0.3930 & -0.0242 & 0 & 0 & 0 & 0 & 0 \\ 0.0494 & -0.2620 & 0.0161 & 0 & 0 & 0 & 0 & 0 \\ 0.1481 & 0.2140 & 0.0484 & 0 & 0 & 0 & 0 & 0 \\ -0.0846 & -0.1223 & 0.1152 & 0 & 0 & 0 & 0 & 0 \\ \hline 0 & 0 & 0 & 0.1250 & 0 & 0 & 0 & 0 \\ 0 & 0 & 0 & -0.4583 & 0.3333 & 0 & 0 & 0 \\ 0 & 0 & 0 & -0.8250 & 0.6000 & -0.2000 & 0 & 0 \\ 0 & 0 & 0 & 0.7071 & -0.5143 & 0.1714 & 0.1429 & 0 \end{array}\right].$$

Let us take as matrices $\mathcal{A}$ and $\mathcal{B}$ right lower and right upper blocks of the matrix A, respectively, i.e.

$$\mathcal{A} = \begin{bmatrix} 11 & 3 & & \\ & 9 & -5 & \\ & & 6 & 7 \\ & & & 0 \end{bmatrix}, \quad \mathcal{B} = \begin{bmatrix} 0 & 0 & 0 & 0 \\ 0 & 0 & 0 & 0 \\ 0 & 0 & 0 & 0 \\ 8 & 0 & 0 & 0 \end{bmatrix}$$

(see (2.4.2) and (2.4.3)). By formulae (2.6.1), (2.6.2) and (2.6.5), (2.6.6) we obtain

$$Z = \begin{bmatrix} 0 & 0 & 0 & 0 \\ 0.3333 & 0 & 0 & 0 \\ 0.6000 & -0.2000 & 0 & 0 \\ -0.5143 & 0.1714 & 0.1429 & 0 \end{bmatrix}, \quad H = \begin{bmatrix} 0 & 0 & 0 & 0.1250 \\ 0 & 0 & 0 & -0.4583 \\ 0 & 0 & 0 & -0.8250 \\ 0 & 0 & 0 & 0.7071 \end{bmatrix}.$$

The matrices Z and H coincide with right lower and left lower blocks of the matrix A^+, respectively. ◇

2.7 Calculation of the pattern matrices Z and H: type 3

If $m = 1$, then the matrices Z and H, as easily follows from (2.4.2) – (2.4.4), have a simple view:

$$Z = [0]_{1\times 1}, \quad H = \left[0\ldots 0\,\frac{1}{\Delta}\right]_{1\times l}; \tag{2.7.1}$$

notice, that $H = \left[\frac{1}{\Delta}\right]_{1\times 1}$ when $l = 1$.

For $m \geq 2$, the matrix $\mathcal{A}$ is

$$\mathcal{A} = \begin{bmatrix} 0 & b_1 & & & \\ & 0 & b_2 & 0 & \\ & & \ddots & \ddots & \\ & 0 & & 0 & b_{m-1} \\ & & & & 0 \end{bmatrix}$$

(see (2.4.2)). Therefore, taking into account the structure (2.4.30) of the matrix $L(\varepsilon)^{-1}$, we obtain that

$$L(\varepsilon)^{-1}\mathcal{A}^T = \begin{bmatrix} 0 & & & & \\ \dfrac{b_1}{b_1^2+\varepsilon} & 0 & & 0 & \\ & \dfrac{b_2}{b_2^2+\varepsilon} & 0 & & \\ & & \ddots & \ddots & \\ & 0 & & \dfrac{b_{m-1}}{b_{m-1}^2+\varepsilon} & 0 \end{bmatrix}.$$

From this, taking limit as $\varepsilon \to +0$, according to the equality (2.4.31) we get

$$Z = \begin{bmatrix} 0 & & & & \\ b_1^{-1} & 0 & & 0 & \\ & b_2^{-1} & 0 & & \\ & 0 & \ddots & \ddots & \\ & & & b_{m-1}^{-1} & 0 \end{bmatrix}. \tag{2.7.2}$$

Thus, having (2.7.1) and (2.7.2), we formulate the following result.

Lemma 2.7.1 *The entries of the matrix $Z = [z_{ij}]_{m\times m}$ defined in (2.4.31) are as follows:*

1) for $m = 1$:

$$Z = [0]_{1\times 1}; \tag{2.7.3}$$

2) for $m \geq 2$:

$$z_{i\,i-1} = \frac{1}{b_{i-1}}, \quad i = 2, 3, \ldots, m; \qquad z_{ij} = 0, \text{ in other cases.} \tag{2.7.4}$$

Finally, about the matrix H. For $l \geq 2$, as follows from (2.4.3) and (2.4.30),

$$L(\varepsilon)^{-1}\mathcal{B}^T = \begin{bmatrix} 0 & 0 & \dots & 0 & \frac{\Delta}{\Delta^2+\varepsilon} \\ 0 & 0 & \dots & 0 & 0 \\ \vdots & \vdots & \ddots & \vdots & \vdots \\ 0 & 0 & \dots & 0 & 0 \end{bmatrix}.$$

Then, taking limit as $\varepsilon \to +0$, according to the equality (2.4.32) we find

$$H = \begin{bmatrix} 0 & 0 & \dots & 0 & \frac{1}{\Delta} \\ 0 & 0 & \dots & 0 & 0 \\ \vdots & \vdots & \ddots & \vdots & \vdots \\ 0 & 0 & \dots & 0 & 0 \end{bmatrix}. \qquad (2.7.5)$$

Remark 2.7.1 *If $l = 1$, then the matrix H has the following view:*

$$H = \begin{bmatrix} \frac{1}{\Delta} \\ 0 \\ \vdots \\ 0 \end{bmatrix}.$$

Thus, in (2.7.1) and (2.7.5) we have obtained the matrix H, for the values $m = 1$ and $m \geq 2$, respectively. However, it can be easily seen that the view of the matrix given in (2.7.1) is a particular case of (2.7.5).

Lemma 2.7.2 *The entries of the matrix $H = [h_{ij}]_{m\times l}$ defined in (2.4.32) are as follows:*

$$\begin{aligned} h_{1l} &= \frac{1}{\Delta}; \\ h_{ij} &= 0, \text{ in other cases.} \end{aligned} \qquad (2.7.6)$$

Next example illustrates Lemmas 2.7.1 and 2.7.2.

Example 2.7.1 Consider a bidiagonal matrix which is partitioned into blocks as follows:

$$A = \left[\begin{array}{cccc|cccc} 2 & 3 & & & & & & \\ & -3 & 1 & & & & & \\ & & 4 & 7 & & & & \\ & & & 0 & 8 & & & \\ \hline & & & & 0 & 3 & & \\ & & & & & 0 & -5 & \\ & & & & & & 0 & 7 \\ & & & & & & & 0 \end{array}\right].$$

Taking advantage of MATLAB software system, we get

$$A^+ = \left[\begin{array}{cccc|ccc|c} 0.4259 & 0.3930 & -0.0242 & 0 & 0 & 0 & 0 & 0 \\ 0.0494 & -0.2620 & 0.0161 & 0 & 0 & 0 & 0 & 0 \\ 0.1481 & 0.2140 & 0.0484 & 0 & 0 & 0 & 0 & 0 \\ -0.0846 & -0.1223 & 0.1152 & 0 & 0 & 0 & 0 & 0 \\ \hline 0 & 0 & 0 & 0.1250 & 0 & 0 & 0 & 0 \\ 0 & 0 & 0 & 0 & 0.3333 & 0 & 0 & 0 \\ 0 & 0 & 0 & 0 & 0 & -0.2000 & 0 & 0 \\ 0 & 0 & 0 & 0 & 0 & 0 & 0.1429 & 0 \end{array}\right].$$

Let us take as matrices $\mathcal{A}$ and $\mathcal{B}$ right lower and right upper blocks of the matrix A, respectively, i.e.

$$\mathcal{A} = \begin{bmatrix} 0 & 3 & & \\ & 0 & -5 & \\ & & 0 & 7 \\ & & & 0 \end{bmatrix}, \quad \mathcal{B} = \begin{bmatrix} 0 & 0 & 0 & 0 \\ 0 & 0 & 0 & 0 \\ 0 & 0 & 0 & 0 \\ 8 & 0 & 0 & 0 \end{bmatrix}$$

(see (2.4.2) and (2.4.3)). Calculations by formulae (2.7.4) and (2.7.6) give us

$$Z = \begin{bmatrix} 0 & 0 & 0 & 0 \\ 0.3333 & 0 & 0 & 0 \\ 0 & -0.2000 & 0 & 0 \\ 0 & 0 & 0.1429 & 0 \end{bmatrix}, \quad H = \begin{bmatrix} 0 & 0 & 0 & 0.1250 \\ 0 & 0 & 0 & 0 \\ 0 & 0 & 0 & 0 \\ 0 & 0 & 0 & 0 \end{bmatrix}.$$

As we see, the matrices Z and H coincide with right lower and left lower blocks of the matrix A^+, respectively. $\diamond$

2.8 Formulae for the entries of the Moore-Penrose inverse

Based on the results obtained in the previous sections, here we give a closed form expressions for the entries of the Moore-Penrose inverse A^+ of a singular upper bidiagonal matrix A of the form (2.1.1).

To solve the problem for any arrangement of one or more zeros on the main diagonal of the matrix A, in Sections 2.1 and 2.2 we carried out some preliminary constructions, calculations and notations.

At first, in the Section 2.1 we represented the matrix A in the block form (2.1.3), i.e.

$$A = \begin{bmatrix} A_1 & B_1 & & & \\ & A_2 & B_2 & 0 & \\ & & \ddots & \ddots & \\ & 0 & & A_{p-1} & B_{p-1} \\ & & & & A_p \end{bmatrix}.$$

There was specified the structure of the blocks A_k and B_k; ibid the types 1, 2 and 3 of the blocks A_k was identified. By virtue of the partitioning rule, only the last block A_p can be a block of type 1. The blocks B_k are given in (2.1.4).

Next, in the Section 2.2 was shown, that the matrix A^+ has the block form (2.2.5), i.e.

$$A^+ = \begin{bmatrix} Z_1 & & & & \\ H_2 & Z_2 & 0 & & \\ & \ddots & \ddots & & \\ & 0 & H_{p-1} & Z_{p-1} & \\ & & & H_p & Z_p \end{bmatrix},$$

where the blocks Z_k, H_k are defined in accordance with (2.2.2), (2.2.3), (2.2.6) and (2.2.7).

Further, to simplify the record of formulae, in (2.2.11) and (2.2.12) we introduced the local numbering of the nodes in the blocks A_k, $2 \le k \le p$. In this connection, let

$$r_s^{(k)} \equiv \frac{b_s^{(k)}}{d_s^{(k)}},\ s = 1, 2, \ldots, n_k - 1;\ r_0^{(k)} = r_{n_k}^{(k)} = 1 \tag{2.8.1}$$

(see also the notation (2.3.9), for $k = 1$). We recall also the notation (2.1.4) for certain superdiagonal entries of the matrix A, namely,

$$\Delta_k \equiv b_{n_1+n_2+\cdots+n_k},\ 1 \le k \le p-1. \tag{2.8.2}$$

- **Calculation of the blocks Z_k.**

Let us start with the block Z_1. The calculation of this block was accomplished in the Section 2.3. We refer to the Lemma 2.3.8, where formulae for the entries of the block Z_1 was obtained. Notice, that if $p = 1$, i.e. the matrix A comprises only one diagonal block, then obviously $A^+ = Z_1$.

Now let us take up the consideration of the blocks Z_k, $2 \le k \le p$. If A_k is a block of type 2, the formulae for the entries of the block Z_k can be obtained by Lemma 2.6.1, replacing m with n_k and using notation (2.2.12), (2.8.1). Next, if A_k is a block of type 3, then the entries of the block Z_k are derived by Lemma 2.7.1, replacing m with n_k and using notation (2.2.12). As has been said above (see Remark 2.1.1), only the last block A_p in the representation (2.1.3) of the matrix A can be a block of type 1. In this case the entries of the block Z_p are calculated by the formulae obtained in Lemma 2.5.1, replacing m with n_p, Δ with Δ_{p-1} (see (2.8.2)) and taking into account notation (2.2.12), (2.8.1).

Thus we arrive at the following statement.

Theorem 2.8.1 *Let a singular upper bidiagonal matrix A given in (2.1.1) with nonzero superdiagonal entries is represented in the block form (2.1.3), according to the rule described in the Section 2.1. Then the entries of diagonal blocks $Z_k = [z_{ij}^{(k)}]_{n_k \times n_k}$, $1 \le k \le p$ in the block representation (2.2.5) of the matrix A^+ are calculated as follows.*

I. The entries of the block Z_1:

1) if A_1 is a block of type 2, then

1a) for indices $j = 1, 2, \dots, n_1 - 1$ *and* $i = 1, 2, \dots, j$:

$$z_{ij}^{(1)} = \frac{(-1)^{i+j} \sum\limits_{k=1}^{n_1-j} \left(\prod\limits_{s=j}^{n_1-k} \frac{1}{r_s^{(1)}} \right) \left(\prod\limits_{s=n_1-k+1}^{n_1-1} r_s^{(1)} \right)}{\prod\limits_{s=1}^{i-1} r_s^{(1)} \cdot d_j \sum\limits_{k=1}^{n_1} \left(\prod\limits_{s=1}^{n_1-k} \frac{1}{r_s^{(1)}} \right) \left(\prod\limits_{s=n_1-k+1}^{n_1-1} r_s^{(1)} \right)}; \tag{2.8.3}$$

1b) for indices $j = 1, 2, \dots, n_1 - 1$ *and* $i = j+1, j+2, \dots, n_1$:

$$z_{ij}^{(1)} = \frac{(-1)^{i+j+1} \left(\prod\limits_{s=i}^{n_1-1} r_s^{(1)} \right) \sum\limits_{k=1}^{j} \left(\prod\limits_{s=1}^{k-1} \frac{1}{r_s^{(1)}} \right) \left(\prod\limits_{s=k}^{j-1} r_s^{(1)} \right)}{d_j \sum\limits_{k=1}^{n_1} \left(\prod\limits_{s=1}^{n_1-k} \frac{1}{r_s^{(1)}} \right) \left(\prod\limits_{s=n_1-k+1}^{n_1-1} r_s^{(1)} \right)}; \tag{2.8.4}$$

1c) for index $j = n_1$:

$$z_{i\,n_1}^{(1)} = 0, \quad i = 1, 2, \dots, n_1; \tag{2.8.5}$$

2) if A_1 *is a block of type 3, then*

2a) for $n_1 = 1$:

$$Z_1 = [0]_{1\times 1}; \tag{2.8.6}$$

2b) for $n_1 \geq 2$:

$$\begin{aligned} z_{i\,i-1}^{(1)} &= \frac{1}{b_{i-1}}, \; i = 2, 3, \dots, n_1; \\ z_{ij}^{(1)} &= 0, \text{ in other cases.} \end{aligned} \tag{2.8.7}$$

II. The entries of the blocks Z_k, $2 \leq k \leq p$:

3) if A_k *is a block of type 2, then*

3a) for indices $j = 1, 2, \dots, n_k$ *and* $i = 1, 2, \dots, j$:

$$z_{ij}^{(k)} = 0; \tag{2.8.8}$$

3b) for indices $j = 1, 2, \dots, n_k - 1$ *and* $i = j+1, j+2, \dots, n_k$:

$$z_{ij}^{(k)} = \frac{(-1)^{i+j+1}}{d_j^{(k)}} \prod_{s=j}^{i-1} \frac{1}{r_s^{(k)}}; \tag{2.8.9}$$

4) if A_k *is a block of type 3, then*

4a) for $n_k = 1$:

$$Z_k = [0]_{1\times 1}; \tag{2.8.10}$$

4b) for $n_k \geq 2$:

$$z_{i\,i-1}^{(k)} = \frac{1}{b_{i-1}^{(k)}}, \; i = 2, 3, \ldots, n_k\,; \qquad z_{ij}^{(k)} = 0, \; \textit{in other cases}; \tag{2.8.11}$$

5) if A_p is a block of type 1, then

5a) for indices $j = 1, 2, \ldots, n_p$ and $i = 1, 2, \ldots, j$:

$$z_{ij}^{(p)} = \frac{(-1)^{i+j}\Delta_{p-1}^2 \sum\limits_{k=0}^{i-1} \frac{1}{b_k^{(p)2}} \left(\prod\limits_{s=1}^{k} r_s^{(p)}\right)\left(\prod\limits_{s=k+1}^{i-1} \frac{1}{r_s^{(p)}}\right)}{d_j^{(p)}\left(\prod\limits_{s=j}^{n_p-1} r_s^{(p)}\right)\left[\prod\limits_{s=1}^{n_p-1}\frac{1}{r_s^{(p)}} + \Delta_{p-1}^2 \sum\limits_{k=1}^{n_p}\frac{1}{d_k^{(p)2}}\left(\prod\limits_{s=1}^{k-1} r_s^{(p)}\right)\left(\prod\limits_{s=k}^{n_p-1}\frac{1}{r_s^{(p)}}\right)\right]}\,; \tag{2.8.12}$$

5b) for indices $j = 1, 2, \ldots, n_p - 1$ and $i = j+1, j+2, \ldots, n_p$:

$$z_{ij}^{(p)} = \frac{(-1)^{i+j+1}\Delta_{p-1}^2 \left(\prod\limits_{s=1}^{j} r_s^{(p)}\right) \sum\limits_{k=i}^{n_p} \frac{1}{d_k^{(p)2}} \left(\prod\limits_{s=i}^{k-1} r_s^{(p)}\right)\left(\prod\limits_{s=k}^{n_p-1} \frac{1}{r_s^{(p)}}\right)}{b_j^{(p)}\left[\prod\limits_{s=1}^{n_p-1}\frac{1}{r_s^{(p)}} + \Delta_{p-1}^2 \sum\limits_{k=1}^{n_p}\frac{1}{d_k^{(p)2}}\left(\prod\limits_{s=1}^{k-1} r_s^{(p)}\right)\left(\prod\limits_{s=k}^{n_p-1}\frac{1}{r_s^{(p)}}\right)\right]}. \tag{2.8.13}$$

- **Calculation of the blocks** H_k.

Let us proceed to considering the blocks H_k, $2 \leq k \leq p$ in the block representation (2.2.5) of the matrix A^+. If A_k is a block of type 2, then the entries of the corresponding block H_k are calculated by the formulae derived in Lemma 2.6.2, replacing m with n_k, l with n_{k-1}, Δ with Δ_{k-1} defined in (2.8.2) and using notation (2.2.12), (2.8.1). Next, if A_k is a block of type 3, the entries of the block H_k are calculated according to Lemma 2.7.2, replacing m with n_k, l with n_{k-1} and Δ with Δ_{k-1}. Finally, if A_p is a block of type 1, then the entries of the block H_p are calculated by the formulae obtained in Lemma 2.5.2, replacing m with n_p, l with n_{p-1}, Δ with Δ_{p-1} and using notation (2.2.12), (2.8.1).

As a result we get the following statement.

Theorem 2.8.2 *Let a singular upper bidiagonal matrix A given in (2.1.1) with nonzero superdiagonal entries is represented in the block form (2.1.3), according to the rule described in the Section 2.1. Then the entries of subdiagonal blocks $H_k = [h_{ij}^{(k)}]_{n_k \times n_{k-1}}$, $2 \leq k \leq p$ in the block representation (2.2.5) of the matrix A^+ are calculated as follows:*

1) if A_k is a block of type 2, then

$$h_{i\,n_{k-1}}^{(k)} = \frac{(-1)^{i+1}}{\Delta_{k-1}} \prod_{s=1}^{i-1} \frac{1}{r_s^{(k)}}, \; i = 1, 2, \ldots, n_k\,; \qquad h_{ij}^{(k)} = 0, \; \textit{in other cases}; \tag{2.8.14}$$

2) if A_k is a block of type 3, then

$$h^{(k)}_{1\,n_{k-1}} = \frac{1}{\Delta_{k-1}}\,; \qquad (2.8.15)$$

$$h^{(k)}_{ij} = 0, \text{ in other cases;}$$

3) if A_p is a block of type 1, then

$$h^{(p)}_{i\,n_{k-1}} = \frac{(-1)^{i+1}\,\Delta_{p-1}\sum\limits_{k=i}^{n_p}\frac{1}{d_k^{(p)2}}\left(\prod\limits_{s=i}^{k-1} r_s^{(p)}\right)\left(\prod\limits_{s=k}^{n_p-1}\frac{1}{r_s^{(p)}}\right)}{\prod\limits_{s=1}^{n_p-1}\frac{1}{r_s^{(p)}} + \Delta^2_{p-1}\sum\limits_{k=1}^{n_p}\frac{1}{d_k^{(p)2}}\left(\prod\limits_{s=1}^{k-1} r_s^{(p)}\right)\left(\prod\limits_{s=k}^{n_p-1}\frac{1}{r_s^{(p)}}\right)}, \qquad (2.8.16)$$

$$i = 1, 2, \ldots, n_k\,;$$

$h^{(p)}_{ij} = 0$, *in other cases.*

Thus in Theorems 2.8.1 and 2.8.2 we have obtained closed form expressions for the entries of the Moore-Penrose inverse of singular upper bidiagonal matrices. In the next Chapter 3 we discuss an issue of practical computation of the matrix A^+.

Chapter 3

AN ALGORITHM TO COMPUTE THE MOORE-PENROSE INVERSE

Based on the formulae and recurrence relations obtained in Chapter 2, here we suggest a fairly simple numerical algorithm to compute the entries of the Moore-Penrose inverse A^+ for a singular upper bidiagonal matrix A of the form (2.1.1).

Recall that to solve the problem, we represented the matrix A in the block form (2.1.3), according to the rule described in the Section 2.1. Using the block structure of the matrix, in the Section 2.2 was shown that the matrix A^+ has the block form (2.2.5) with diagonal blocks Z_k and subdiagonal blocks H_k. Let us turn to the description of algorithms for computing these blocks.

3.1 An algorithm of computing the block Z_1

Summarizing the considerations of the Section 2.3, let us write the following procedure to compute the entries of the first diagonal block $Z_1 = \left[z_{ij}^{(1)}\right]_{n_1 \times n_1}$.

Algorithm Z1 $(A_1, n_1 \Rightarrow Z_1)$

1. If A_1 is a block of type 2, then:

   1.1 Compute the quantities $r_s^{(1)}$ (see (2.3.9)):

$$r_s^{(1)} = \frac{b_s}{d_s}, \quad s = 1, 2, \ldots, n_1 - 1; \; r_0^{(1)} = r_{n_1}^{(1)} = 1 .$$

   1.2 Compute the quantities $\overset{\circ}{\mu}_i$ (see (2.3.7), (2.3.11)):

$$\overset{\circ}{\mu}_{n_1} = 1; \quad \overset{\circ}{\mu}_i = -r_i^{(1)} \, \overset{\circ}{\mu}_{i+1}, \; i = n_1 - 1, n_1 - 2, \ldots, 1 .$$

   1.3 Compute the quantities $\overset{\circ}{\nu}_i$ (see (2.3.13), (2.3.16)):

$$\overset{\circ}{\nu}_1 = 1; \quad \overset{\circ}{\nu}_{i+1} = -\frac{1}{r_i^{(1)}} \, \overset{\circ}{\nu}_i, \; i = 1, 2, \ldots, n_1 - 1 .$$

1.4 Compute the quantities u_j (see (2.3.29)):

$$u_{n_1-1} = -\frac{1}{b_{n_1-1}};$$

$$u_j = -\frac{d_{j+1}u_{j+1} + \overset{\circ}{\mu}_{j+1}}{b_j}, \quad j = n_1 - 2, n_1 - 3, \ldots, 1.$$

1.5 Compute the quantities w_j (see (2.3.39)):

$$w_1 = -\frac{1}{d_1};$$

$$w_j = -\frac{b_{j-1}w_{j-1} + \overset{\circ}{\nu}_j}{d_j}, \quad j = 2, 3, \ldots, n_1 - 1.$$

1.6 Compute the quantity q (see (2.3.26)):

$$q = \overset{\circ}{\mu}_1 + d_1 u_1 .$$

1.7 Zeroing the last column of the block Z_1 (see (2.3.20)):

$$z^{(1)}_{i\,n_1} = 0\,, \quad i = 1, 2, \ldots, n_1 .$$

1.8 Compute the upper triangular part of the block Z_1 (see (2.3.25)):

$$z^{(1)}_{ij} = \frac{\overset{\circ}{\nu}_i\, u_j}{q}, \quad i = 1, 2, \ldots, j; \; j = 1, 2, \ldots, n_1 - 1.$$

1.9 Compute the lower triangular part of the block Z_1 (see (2.3.36)):

$$z^{(1)}_{ij} = \frac{\overset{\circ}{\mu}_i\, w_j}{q}, \quad i = j+1, j+2, \ldots, n_1; \; j = 1, 2, \ldots, n_1 - 1.$$

2. If A_1 is a block of type 3, then:

2.1 If $n_1 = 1$ (see (2.3.47)): $Z_1 = [0]_{1\times 1}$.

2.2 If $n_1 \geq 2$ (see (2.3.48)):

$$z^{(1)}_{i\,i-1} = \frac{1}{b_{i-1}}, \quad i = 2, 3, \ldots, n_1;$$

$$z^{(1)}_{ij} = 0, \text{ in other cases.}$$

End algorithm

Direct calculations show that the numerical implementation of the **Algorithm Z1** requires no more than

$$A_{ops}(Z_1) = n_1^2 + O(n_1) \tag{3.1.1}$$

arithmetical operations.

3.2 Algorithms for the pattern matrices Z and H

Having intermediate formulae and recurrence relations revealed in the Sections 2.4 – 2.7, below we propose numerical algorithms to compute the entries of the pattern matrices $Z = [z_{ij}]_{m\times m}$ and $H = [h_{ij}]_{m\times l}$, in accordance with the types 1, 2 and 3 specified in the Section 2.4.

Algorithm Z, H/type 1 $(\mathcal{A}, \Delta, m, l \Rightarrow Z, H)$

1. Compute the quantities r_s (see (2.4.5)):

$$r_s = \frac{b_s}{d_s}, \; s = 1, 2, \dots, m-1; \; r_0 = r_m = 1 .$$

2. Compute the quantities α_i (see (2.4.16)):

$$\alpha_m = 1; \quad \alpha_i = -r_i \alpha_{i+1}, \; i = m-1, m-2, \dots, 1 .$$

3. Compute the quantities β_i (see (2.4.22)):

$$\beta_1 = -r_1; \quad \beta_{i+1} = -r_{i+1}\beta_i, \; i = 1, 2, \dots, m-2 .$$

4. Compute the quantities $\overset{\circ}{\mu}_i$ (see (2.4.12), (2.4.14)):

$$\overset{\circ}{\mu}_m = 1; \quad \overset{\circ}{\mu}_i = -r_i \overset{\circ}{\mu}_{i+1} + \frac{d_m^2}{d_i^2}\frac{1}{\alpha_i}, \; i = m-1, m-2, \dots, 1 .$$

5. Compute the quantities $\overset{\circ}{\nu}_i$ (see (2.4.18), (2.4.20)):

$$\overset{\circ}{\nu}_1 = 1; \quad \overset{\circ}{\nu}_{i+1} = -\frac{1}{r_i}\overset{\circ}{\nu}_i + \frac{\Delta^2}{b_i^2}\beta_i, \; i = 1, 2, \dots, m-1 .$$

6. Compute the quantity q (see (2.4.24)):

$$q = \frac{d_m^2}{\alpha_1} + \Delta^2 \overset{\circ}{\mu}_1 .$$

7. Compute the upper triangular part of the matrix Z (see (2.5.6)):

$$z_{ij} = \overset{\circ}{\nu}_i \frac{1}{q}\frac{d_m^2}{d_j}\frac{1}{\alpha_j}, \; i = 1, 2, \dots, j; \; j = 1, 2, \dots, m .$$

8. Compute the lower triangular part of the matrix Z (see (2.5.9)):

$$z_{ij} = \overset{\circ}{\mu}_i \frac{1}{q}\frac{\Delta^2}{b_j}\beta_j, \; i = j+1, j+2, \dots, m; \; j = 1, 2, \dots, m-1 .$$

9. Zeroing the first $l-1$ columns of the matrix H (see (2.5.20)):

$$h_{ij} = 0, \; i = 1, 2, \dots, m; \; j = 1, 2, \dots, l-1 .$$

10. Compute the last column of the matrix H (see (2.5.18)):

$$h_{il} = \overset{\circ}{\mu}_i \frac{\Delta}{q}, \; i = 1, 2, \ldots, m.$$

End algorithm

By direct calculations we find that the **Algorithm Z, H/type 1** requires

$$A_{ops}(Z, H/1) = m^2 + O(m) \tag{3.2.1}$$

arithmetical operations.

Algorithm Z, H/type 2 $(\mathcal{A}, \Delta, m, l \Rightarrow Z, H)$

1. Compute the quantities r_s (see (2.4.5)):

$$r_s = \frac{b_s}{d_s}, \; s = 1, 2, \ldots, m-1; \; r_0 = r_m = 1.$$

2. Zeroing the upper triangular part of the matrix Z (see (2.6.1)):

$$z_{ij} = 0, \; i = 1, 2, \ldots, j; \; j = 1, 2, \ldots, m.$$

3. Compute the lower triangular part of the matrix Z (see (2.6.3), (2.6.4)):

$$z_{j+1\,j} = \frac{1}{b_j}; \; z_{ij} = -\frac{z_{i-1\,j}}{r_{i-1}}, \; i = j+2, j+3, \ldots, m;$$
$$j = 1, 2, \ldots, m-1.$$

4. Zeroing the first $l-1$ columns of the matrix H (see (2.6.5)):

$$h_{ij} = 0, \; i = 1, 2, \ldots, m; \; j = 1, 2, \ldots, l-1.$$

5. Compute the last column of the matrix H (see (2.6.7), (2.6.8)):

$$h_{1l} = \frac{1}{\Delta}; \; h_{il} = -\frac{h_{i-1\,l}}{r_{i-1}}, \; i = 2, 3, \ldots, m.$$

End algorithm

The numerical implementation of the **Algorithm Z, H/type 2** requires

$$A_{ops}(Z, H/2) = \frac{1}{2}m^2 + O(m) \tag{3.2.2}$$

arithmetical operations.

Algorithm Z, H/type 3 $(\mathcal{A}, \Delta, m, l \Rightarrow Z, H)$

1. Compute the entries of the matrix Z (see (2.7.3), (2.7.4)):

   1.1 If $m = 1$, then $Z = [0]_{1\times 1}$.

   1.2 If $m \geq 2$, then

$$z_{i\,i-1} = \frac{1}{b_{i-1}}, \quad i = 2, 3, \ldots, m;$$
$$z_{ij} = 0, \text{ in other cases.}$$

2. Compute the entries of the matrix H (see (2.7.6)):

$$h_{1\,l} = \frac{1}{\Delta};$$
$$h_{ij} = 0, \text{ in other cases.}$$

End algorithm

Obviously that the **Algorithm Z, H/type 3** requires

$$A_{ops}(Z, H/3) = m \tag{3.2.3}$$

arithmetical operations.

3.3 A general computational procedure

Taking advantage of the numerical algorithms developed in the Sections 3.1 and 3.2, below we give a procedure to compute the Moore-Penrose inverse for singular upper bidiagonal matrices.

Procedure 2D/MPinv $(A, n \Rightarrow A^+)$

Input: an upper bidiagonal matrix A of the form (2.1.1).

1. A partition (2.1.3) of the matrix A into blocks, according to the rule specified in the Section 2.1; identification of the blocks A_k $(1 \leq k \leq p)$, B_k $(1 \leq k \leq p-1)$ and determination of the parameters n_k $(1 \leq k \leq p)$ which define the block sizes. In each block A_k, $2 \leq k \leq p$, the local numbering (2.2.11), (2.2.12) of the entries is introduced. Input of the quantities Δ_k, $1 \leq k \leq p-1$ (see (2.8.2)).

2. The block Z_1 in the block representation (2.2.5) of the matrix A^+ is computed. For that the **Algorithm Z1** $(A_1, n_1 \Rightarrow Z_1)$ is used. The algorithm requires no more than

$$A_{ops}(Z_1) = n_1^2 + O(n_1) \tag{3.3.1}$$

   arithmetical operations (see (3.1.1)).

   If $p = 1$, the computations are completed and $A^+ = Z_1$. Otherwise, if $p \geq 2$, then proceed to successive computation of the blocks Z_k and H_k.

3. Computation of the blocks Z_k, H_k, for the values $k = 2, 3, \ldots, p$.

   (a) If A_k is a block of type 2, the blocks Z_k and H_k are computed using the **Algorithm Z, H/type 2** $(A_k, \Delta_{k-1}, n_k, n_{k-1} \Rightarrow Z_k, H_k)$. The algorithm requires

$$A_{ops}(Z_k, H_k/2) = \frac{1}{2}n_k^2 + O(n_k) \tag{3.3.2}$$

   arithmetical operations (see (3.2.2)).

   (b) If A_k is a block of type 3, the blocks Z_k and H_k are computed by the **Algorithm Z, H/type 3** $(A_k, \Delta_{k-1}, n_k, n_{k-1} \Rightarrow Z_k, H_k)$. The algorithm requires

$$A_{ops}(Z_k, H_k/3) = n_k \tag{3.3.3}$$

   arithmetical operations (see (3.2.3)).

   (c) If A_p is a block of type 1, the blocks Z_p and H_p are computed using the **Algorithm Z, H/type 1** $(A_p, \Delta_{p-1}, n_p, n_{p-1} \Rightarrow Z_p, H_p)$. The algorithm requires

$$A_{ops}(Z_p, H_p/1) = n_p^2 + O(n_p) \tag{3.3.4}$$

   arithmetical operations (see (3.2.1)).

Output: matrix A^+ obtained in the block form (2.2.5).

End procedure

It may easily make sure that the described computational procedure requires no more than

$$A_{ops}(2D/MPinv) = n_1^2 + \frac{1}{2}\sum_{k=2}^{p-1} n_k^2 + n_p^2 + O(n) \tag{3.3.5}$$

arithmetical operations (recall that $n_1 + n_2 + \cdots + n_p = n$).

Thus, we can formulate the following statement.

Proposition 3.3.1 *Let A be a singular upper bidiagonal matrix of the form (2.1.1) with nonzero superdiagonal entries. Then the Moore-Penrose inverse A^+ of this matrix can be computed using the* **Procedure 2D/MPinv** *with an expenditure of $A_{ops}(2D/MPinv)$ arithmetical operations estimated in (3.3.5).*

We emphasize the following important feature of the **Procedure 2D/MPinv**. Proceeding from the structure of the blocks in the block representation (2.2.5) of the matrix A^+ (namely, the presence of zeros located at predetermined places) and the estimations of the number of arithmetical operations required to compute each block (see (3.3.1) – (3.3.4)), we can assert that for computing one nonzero entry of the matrix A^+ asymptotically one arithmetical operation is expended. Thereby the proposed computational method can be considered as optimal.

Below we give an example to illustrate the work of the computational procedure.

Example 3.3.1 Consider a matrix, which is divided into blocks as follows:

$$A = \left[\begin{array}{ccc|ccc|c|ccc}
2 & 5 & & & & & & & & \\
 & 3 & -7 & & & & & & & \\
 & & 0 & 6 & & & & & & \\
\hline
 & & & 4 & 2 & & & & & \\
 & & & & -5 & -1 & & & & \\
 & & & & & 0 & 4 & & & \\
\hline
 & & & & & & 0 & 2 & & \\
\hline
 & & & & & & & 3 & -4 & \\
 & & & & & & & & -6 & 3 \\
 & & & & & & & & & 8
\end{array}\right].$$

Applying the **Procedure 2D/MPinv**, we get

$$A^+ = \left[\begin{array}{ccc|ccc|c|ccc}
0.0796 & -0.0206 & 0.0000 & & & & & & & \\
0.1682 & 0.0082 & 0.0000 & & & & & & & \\
0.0721 & -0.1393 & 0.0000 & & & & & & & \\
\hline
0.0000 & 0.0000 & 0.1667 & 0.0000 & 0.0000 & 0.0000 & & & & \\
0.0000 & 0.0000 & -0.3333 & 0.5000 & 0.0000 & 0.0000 & & & & \\
0.0000 & 0.0000 & 1.6667 & -2.5000 & -1.0000 & 0.0000 & & & & \\
\hline
 & & & 0.0000 & 0.0000 & 0.2500 & 0.0000 & & & \\
\hline
 & & & & & & 0.2006 & 0.1996 & -0.1331 & 0.0499 \\
 & & & & & & 0.0506 & -0.0337 & -0.1442 & 0.0541 \\
 & & & & & & 0.0125 & -0.0083 & 0.0055 & 0.1229
\end{array}\right].$$

The result coincides with the computations done with MATLAB. $\diamond$

In conclusion, we point out another important property of the computational procedure. Each pair of blocks Z_k, H_k in the block representation (2.2.5) of the matrix A^+, for different values of k, is computed independently of each other. From this point of view, the algorithm can be effectively implemented on computers with parallel architecture.

Bibliography

[1] Albert A. Regression and the Moore-Penrose Pseudoinverse. Academic Press, New York, 1972.

[2] Ben-Israel A. and T.N.E. Greville. Generalized Inverses. Theory and Applications, 2nd ed.. Springer, New York, 2003.

[3] Bjerhammar A. Application of calculus of matrices to method of least squares with special reference to geodetic calculations. *Trans. Roy. Inst. Tech.*, Stockholm, No.49, 1951, pp. 89.

[4] Bjerhammar A. Rectangular reciprocal matrices, with special reference to geodetic calculations. *Bull. Géodésique*, 1951, 188-220.

[5] Fredholm I. Sur une class d'équations fonctionnelles. *Acta Math.*, **27**, 1903, 365-390.

[6] Golub G.H. and W. Kahan. Calculating the Singular Values and Pseudo-Inverse of a Matrix. *SIAM J. Num. Anal.*, **2**, 1965, 205-224.

[7] Golub G.H. and Ch.F. van Loan. Matrix Computations, 3rd ed.. The John Hopkins University Press, 1996.

[8] Golub G.H. and C.Reinsch. Singular Value Decomposition and Least Squares Solutions. *Numer. Math.*, **14**, 1970, 403-420.

[9] Hakopian Yu.R. and S.S.Aleksanyan. Moore-Penrose inverse of bidiagonal matrices. I. *Proceedings of the Yerevan State University*, **2**, 2015, 11-20.

[10] Hakopian Yu.R. and S.S.Aleksanyan. Moore-Penrose inverse of bidiagonal matrices. II. *Proceedings of the Yerevan State University*, **3**, 2015, 8-16.

[11] Hakopian Yu.R. and S.S.Aleksanyan. Moore-Penrose inverse of bidiagonal matrices. III. *Proceedings of the Yerevan State University*, **1**, 2016, 12-21.

[12] Hakopian Yu.R. and S.S.Aleksanyan. Moore-Penrose inverse of bidiagonal matrices. IV. *Proceedings of the Yerevan State University*, **2**, 2016, 28-34.

[13] Hurwitz W.A. On the pseudo-resolvent to the kernel of an integral equation. *Trans. Amer. Math. Soc.*, **13**, 1912, 405-418.

[14] Lawson C.L. and R.J. Hanson. Solving Least Squares Problems. Prentice-Hall Inc., Englewood Cliffs, N.J., 1974.

[15] Lewis J.W. Invertion of tridiagonal matrices. *Numer. Math.*, **38**, 1982, 333-345.

[16] Moore E. On the reciprocal of the general algebraic matrix. *Bull. Amer. Math. Soc.*, **26**, 1920, 394-395.

[17] Penrose R. A generalized inverse for matrices. *Proc. Cambridge Philos. Soc.*, **51**, 1955, 406-413.

www.ingramcontent.com/pod-product-compliance
Lightning Source LLC
Chambersburg PA
CBHW081126180526
45170CB00008B/3027